MOBILE COMMUNICATIONS

WILEY SERIES IN COMMUNICATION AND DISTRIBUTED SYSTEMS

MOBILE COMMUNICATIONS

A. Jagoda

and

M. de Villepin

Dauphine University, Paris
France

Translated by J. C. C. Nelson

University of Leeds
UK

JOHN WILEY & SONS
Chichester · New York · Brisbane · Toronto · Singapore

First published under the title *Les télécommunications mobiles* by Editions Eyrolles,
© Eyrolles, Paris, 1991

Copyright © 1993 by John Wiley & Sons Ltd,
 Baffins Lane, Chichester,
 West Sussex PO19 1UD, England

Other Wiley Editorial Offices

John Wiley & Sons, Inc., 605 Third Avenue,
New York, NY 10158-0012, USA

Jacaranda Wiley Ltd, G.P.O. Box 859, Brisbane,
Queensland 4001, Australia

John Wiley & Sons (Canada) Ltd, 22 Worcester Road,
Rexdale, Ontario M9W 1L1, Canada

John Wiley & Sons (SEA) Pte Ltd, 37 Jalan Pemimpin #05-04,
Block B, Union Industrial Building, Singapore 2057

Library of Congress Cataloging-in-Publication Data

Jagoda, A.
  [Télécommunications mobiles. English]
  Mobile communications / A. Jagoda, M. de Villepin : translated by
J.C.C. Nelson.
     p. cm. — (Wiley series in communication and distributed
systems)
  Includes index.
  ISBN 0 471 93906 4
  1. Mobile communication systems. I. Villepin, M. de. II. Title.
III. Series.
TK6570.M3J33 1993
384.5′3′094—dc20   93-9127
                                                                       CIP

British Library Cataloguing in Publication Data

A catalogue record for this book is available from the British Library

ISBN 0 471 93906 4

Typeset by Mathematical Composition Setters Ltd, Salisbury, Wiltshire
Printed and bound in Great Britain by Biddles Ltd, Guildford and Kings Lynn

CONTENTS

PART 1 MOBILE TELECOMMUNICATIONS

PART 2 MOBILE SERVICES AND PRODUCTS

Part 1

MOBILE TELECOMMUNICATIONS

1 INTRODUCTION

1.1 COMMUNICATION WHILE TRAVELLING

The pace of our daily life has been increasing for several decades. Our needs have multiplied as new products have appeared and then been replaced after a few years, or even months, of existence by a more fashionable product or one of higher performance. The life cycles of the technologies used in consumer and professional electronic products are also becoming shorter. This acceleration is an inherent fact of our consumer society.

Lifestyles and the relationship between people and machines are changing due to the multiplicity of ephemeral consumer products. Objects no longer have a history; they are merely tools which fulfil a predetermined function.

Personal portable products are of a new type which has appeared among these impersonal objects. This category includes watches, pens, wallets, handbags, calculators, portable radios and pocket telephones.

As these products for the pocket are carried on one's person, they belong in a very personal way and have, therefore, a specific identity corresponding to the image which they are given.

In the evolution of lifestyles, the explosive increase of travel and time management are major factors. The pocket telephone is, therefore, remarkable for two reasons. It is not only an impersonal tool or product but is also very much a personal portable product. The possibility of distant communication while travelling, being able to call or be called at will anywhere at any time permits the pocket telephone to be often considered as a desirable, almost magic, personal item.

The end of the twentieth century will see the appearance, as stated by Jacques Attali in *Lignes d'horizon*[1], of mobile devices; the pocket telephone is one example of these. These next ten years will see the explosive growth

---

[1] Fayard (1990).

of mobile services which affect the general public. This book, therefore, attempts to define the evolution of current and future mobile services which have led, or will lead to, the birth of the pocket telephone market for both the professional and general public.

1.2   COVERAGE AND OBJECTIVE

This book is concerned with all civil mobile telecommunication services and products. Radio and satellite broadcasting systems and mobile systems for military use are intentionally not covered.

To analyse the current position of mobile products and services in Europe and the developments in this area during the next few years with the emergence of European telepoint, paging and digital radiotelephony services, some basic technical aspects of these systems will be presented in the first part of the book. The statutory aspects associated with the provision of European telecommunication standards will also be addressed.

The second part includes an exhaustive analysis of civil mobile services, including the outlook and strategy of those involved in this sector—standards organisations, operators, manufacturers, distributors and users of mobile telecommunication services and products.

The third and last part attempts to describe the foreseeable developments of this changing sector to the year 2000. The emphasis is on the birth of the pocket telephone and its industrial and political implications for the European economy in the face of world markets.

The object of this work is to present, thoroughly analyse and assess the prospects for the mobile civil communication sector at European level together with an overview of those involved. The ideas and opinions presented in this book are personal to the authors.

The book is not, therefore, restricted to communications professionals; it is for all who wish to know more about their telephone and seek a solution to their mobile communication requirements.

1.3   MODE OF USE

To assist the novice reader, the book has been planned with several entry points. Reading is not necessarily sequential, although unity is provided by three linked parts. These parts can initially be considered to be independent. The work may be approached as follows.

The first part may be reread in order to grasp the implications of the technical and statutory constraints on the development of current and future mobile services as described in the second and third parts.

Similarly, the chapters in the second part can be tackled separately according to the needs or interest of the reader. Each chapter covers one particular type of mobile service (radio paging, private radio communication, cellular radio communication, cordless telephony, marine, satellite and aircraft communication) which is treated comprehensively and independently.

The third part presents the evolution of mobile services and their economic and political implications. This can also be tackled separately from the rest of the work but it does require previous knowledge of mobile telecommunications.

1.4   HISTORY AND IMPORTANCE

In the next ten years, mobile communications will experience an upheaval comparable to that experienced by aviation and the motor car since the start of the century, and computing since 1950.

Mobile communications using private networks remained in their infancy until 1970. Subsequently it has seen a large increase with the introduction of radio telephone networks and cordless telephones. In the next ten years there will be an explosion of mobile services under the combined pressures of technological progress in information processing and strong market demand.

Experts predict that in the year 2000, from 20 to 40% of telecommunications operators' revenue will be from mobile services with fifty million subscribers in Europe and one telephone in two will be cordless.

1.5   PROSPECTS FOR THE EUROPEAN POCKET TELEPHONE

There are four aspects to the European pocket telephone; these are technological, economic, manufacturing and sociological.

In the technological area, finalisation, validation and establishment of European standards to permit the installation of easily usable infrastructures and miniaturised low cost terminals are being realised for various mobile radio communication services and products; this was a very ambitious gamble initiated in the 1980s.

In the economic area, the pocket telephone represents a considerable opportunity: of the order of 60 thousand million ecus per year is envisaged for products and services by the year 2000. It is the fastest growing telecommunication sector, of the order of 20% per year.

In the manufacturing area, the pocket telephone is an opportunity for European industry provided that European regulations and initiatives are put in place. A lack of short term vigilance would be disastrous for the European terminal industry whose main weakness is associated with the European electronic component industry.

Finally, from the sociological point of view, our way of life will be made easier for some and restricted for others by the portable pocket telephone. Previously limited to public authorities, government and the armed forces, the cordless telephone became available to the professional and the general public in the years between 1970 and 1980. In its pocket version, the telephone becomes multidimensional, both a tool bringing knowledge and power to the bearer and a fashionable, fun object, indicative of our ubiquity.

2 TECHNICAL CONCEPTS

2.1 DEFINITIONS

2.1.1 The radio communication system

There are numerous definitions of telecommunication systems in general and radio communication systems in particular. The most general is a set of facilities which permits terminals to be interconnected via at least one cordless or radio interface, called the air interface. Here, the use of 'terminals' implies radio or other types of communication terminals if communication between a wired terminal, a telephone receiver and a cordless terminal is involved. Fig. 2.1 illustrates this general definition.

This definition applies to the different networks and types of radio communication described in Part 2—personal pagers, private radio communication, cellular radio telephone, cordless telephone systems, mobile communication by satellite and in aircraft. It is clearly evident that the terminals are not part of the network although their specification depends on it, particularly by its physical characteristics and communication protocols used at the air interface.

2.1.2 The constituent elements of a system

There are two classes of network—public networks and private networks. Public radio communication networks are connected to the public communication networks such as the telephone network and other public networks of the X.25 and ISDN type. Private networks are not authorised for connection to the public networks. A public radio communication network is thus often represented as a specific subsection of a more general public communication network.

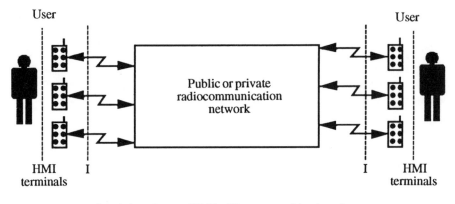

I = air interface – HMI = Human-machine interface

Figure 2.1 General definition of a radio communication system

The radio communication network itself can be subdivided into three subsections or subnetworks: the switching subnetwork, the radio subnetwork and the operating and control subnetwork.

● 1. The function of the switching subnetwork is the routeing of calls. It updates the data concerning users of the service, their subscriptions and their location. It is sometimes integrated into the public network as in the case of radio pagers.

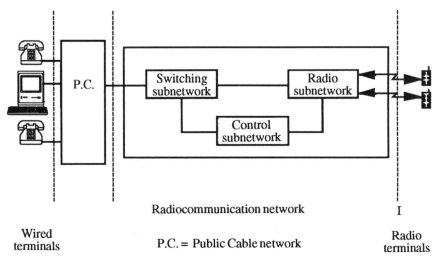

Figure 2.2 The constituent elements of a system

- 2. The function of the radio subnetwork is transmission and reception; it consists of radio equipment called base stations. There can be up to 600 in a national cellular radio telephone network.

- 3. The operating and control subnetwork has the function of supervising and controlling the radio and switching subnetworks. It permits centralised control by the operator.

2.1.3 The constituent elements of a terminal

As with networks, terminals can be divided into subsections: the radio subsection, the logic subsection, the power supply, the user interface and the network interface.

- 1. The functions of the radio subsection include reception and demodulation, modulation and transmission (if provided) and frequency synthesis.

- 2. The radio subsection is controlled by the logic subsection; this includes, in the case of digital terminals, signal processing elements for coding and decoding of the information to be transmitted, encryption, error protection and error correction mechanisms.

- 3. The power supply subsection is sometimes included in the logic subsection; it includes the necessary interfaces with the various types of power supply such as portable batteries, vehicle batteries, the mains and battery chargers.

- 4. The user interface often includes a keyboard and numeric or

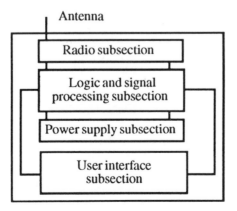

Figure 2.3 The constituent elements of a terminal

alphanumeric display. It is controlled by the 'human–machine interface' software in the logic section.

- 5. An internal or external antenna interfaces with the network.

Fig. 2.3 shows these elements in schematic form.

2.2 THE SPECTRUM AND ITS ALLOCATION

2.2.1 Frequencies

Civil radio communication services use frequency bands situated between several tens of MHz for private systems and several GHz for satellite systems. The processes and the statutory and standardisation organisations involved in the allocation of these frequencies are described in Chapter 4. In particular, in the context of strategic analysis and recommendation, ETSI, the European Telecommunications Standards Institute, proposes the establishment of a permanent organisation at European level for the allocation and optimum use of frequencies for mobile services and radio broadcasting. Furthermore, in view of the substantial efforts to standardise new services and systems, ETSI recommends frequency allocation in the medium and long term.

2.2.2 Mobile services and frequencies

This section (see Fig. 2.4) reviews the frequencies allocated to the principal mobile services described in Part 2 of this book. (Note: for an explanation of the services see the Glossary and Part 2.)

2.3 BASIC TECHNICAL CONCEPTS

2.3.1 Propagation

Propagation is one of the essential elements defining a mobile communication service. For each service, it is studied by operators and manufacturers in order to investigate the received signal quality in urban areas with many obstacles, in rural areas and in buildings. Above 100 MHz, waves have greater difficulty propagating through obstacles and this creates large regions of shadow due to the terrain or buildings. In

Service	Current frequencies	Proposed development
Analogue cellular	200/450/900 MHz bands	Liberation for GSM 890–905/ 935–950 MHz in some countries
Digital cellular		
• GSM	890–915/935–960 MHz	870–888/915–933 MHz
• DCS 1800	1710–1880 MHz	
Personal paging		
• public service	80/160 MHz bands	ERMES 169 MHz
• private site	25, 41/440, 470 MHz	25, 50/146, 174/420, 470 MHz
Cordless telephone		
• analogue		
CT0	16, 26/41, 47 MHz	
CT1	914–915/959–960 MHz	
• digital		
CT2	864–868 MHz	
DECT	1880–1990 MHz	1850–1900 MHz
UMTS cellular services and future cordless		200 MHz to above 1 GHz, of which 140 MHz is for voice and 60 MHz for data
Private radiocommunication		
• conventional PMR networks	30–47/68–87/146–174/ 410–430 MHz 440–470 MHz	47–68/174–230/230–400/ 790–862 MHz
• trunk analogue/digital		230–400/790–862/870–888/ 915–933 MHz
• citizen band	40 frequencies at 27 MHz	
• digital short range radio	933–935/888–890 MHz	
Satellite services	2 MHz band L	10 MHz band L

Figure 2.4 Mobile services and frequency allocation

particular, the horizon considerably reduces the range of transmitted signals due to the large losses in the atmosphere. Conversely, reuse of frequencies beyond the horizon can be achieved without much danger. This is a requirement for the cellular radio telephone, for example.

Modelling is used in an attempt to analyse the relation between the received signal power and the power of the transmitted signal. In practice, a relationship is used in which the received power is related to the transmitted power by a factor which depends on the fourth power of the inverse of the distance. In other words, if receiver A is twice as far from a transmitter as receiver B, the power received by receiver A, all other things being equal, is 16 times less than that received by B. Other factors which determine the received power are the height of the transmitter and receiver above ground level and the shadow created by the terrain and obstacles such as buildings. However, this shadow is reduced by multiple reflections from neighbouring obstacles and consequently the coverage of the received signal, and hence of the service, is increased. In town, direct reception from the transmitter is almost impossible and there are constant multiple trajectory phenomena. When multiple signals are received, these reflections cause the signal which would have been detected at the receiver from a direct trajectory to be either increased or decreased. Very short distances between widely differing maximum and minimum signal levels (from 20–60 dB) are observed.

Propagation into buildings is often required for radio communication services, particularly for personal paging systems. Attenuations of 20 dB are nevertheless observed. The orientation of the receiver with respect to the transmitter can also be a major cause of attenuation with typical variations of 15 dB.

The study of propagation remains a very complex area which is the subject of much modelling and terrain measurements by service operators. In each case 100% coverage and service quality cannot be guaranteed due to propagation problems and the cost of complete coverage particularly in mountainous regions.

2.3.2 Sources of interference and noise

There are many sources of noise and interference which disturb reception of the desired signal.

Interference caused by transmitters, other than the nominal one, which operate on the same frequency is 'co-channel' interference. Interference arising from other transmitters on other frequencies which contribute to the received power is described as 'intermodulation'.

Other sources of noise, for example the electromagnetic environment of a vehicle with respect to an on-board communication terminal, are experienced. Furthermore, independently of the noise sources, the received signal quality can be degraded by poor receiver sensitivity on a particular channel, and interference due to reception on adjacent channels can also arise.

In general, the objective of the system and terminal designer is to limit the total noise generated by the various noise and interference sources to the lowest possible level with respect to the desired signal. In statistical terms, the designer may require, for example, that within the defined coverage of the service, the signal to noise ratio exceeds a predefined value in 97% of cases.

2.3.3 Interference limitation

Most radio communication systems contain only one transmitter. Particularly in cellular networks, efficient spectrum utilisation requires frequency reuse and a large number of transmitters is necessary in order to cover national territory. The existence of several transmitters creates overlapping propagation areas and hence areas of possible interference. There are many methods of minimising interference; the simplest to implement are as follows:

- 1. Frequency separation; this method avoids reuse of the same frequency by neighbouring transmitters or reusing a frequency on the most distant transmitters possible, while making maximum reuse of frequencies.

- 2. Sequential transmission; this method consists of inhibiting the transmission of adjacent transmitters which are likely to cause interference when the main transmitter is active.

2.4 PRINCIPAL SYSTEM CHARACTERISTICS

2.4.1 Systems

There are several forms of network system. They are characterised by a range of criteria which are related to the final service provided for the user. The following are generally included among these criteria: service coverage (urban, regional, national or international); the type of service provided (messages, voice or data), the quality of service provided (network availability, the error rate of the transmitted information, the intercellular transfer time for a cellular radio telephone network and the network access time); the maximum capacity and subscriber density, generally expressed as the number of communications connected per unit frequency per unit area (Erlang/MHz/km^2). These various criteria will be used in Chapter 9 to identify and characterise the various networks.

System definition is more technical and less functional; it concerns the system builder as opposed to the network user. The principal system characteristics generally used relate to the interfaces: the air interface between the system and the terminals; the interface between the system and the external networks to which it can be connected, such as the public telecommunications cable network; and sometimes internal system interfaces between the constituent elements.

2.4.2 Personal paging systems

Personal paging systems are among the simplest mobile systems. The air interface is characterised by the frequency used and the communication protocol. The network interface, in the case of public systems, is provided by the public telephone network or sometimes an X.25 public data network. In the case of a private system, the interface is provided by a business telephone system (PABX) or a local data network. The architecture of a radio paging system is shown in Fig. 2.5.

2.4.3 Cellular systems

These are much more complex than the previous systems and depend on three principles.

- 1. Frequency reuse by distant transmitters, which permits frequency use to be maximised while limiting interference. The service area is thus divided into adjacent cells, each having a transmitter–receiver and a number of frequencies which will be reused in distant cells. The minimum set of cells using all the available frequencies is called a 'cluster' and repeats periodically.

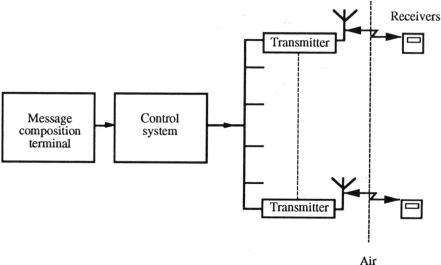

Figure 2.5 General architecture of a personal paging system

- 2. Subdivision of the coverage into cells; this requires a 'handover' mechanism between two adjacent cells in order to ensure continuity of communication for a mobile moving from one cell to another. It depends on a succession of complex operations which include measurement of the signal received by the terminal from the transmitter in the cell and from the transmitters in neighbouring cells, the decision to hand-over following pre-established criteria, execution of the hand-over and updating of the mobile location parameters. Each radio telephone system (see Chapter 6.3) specifies a hand-over time which is generally between 100 and 300 ms.

- 3. To respond to increasing demand from radio telephone users and hence to increase network capacity, the technique of dividing cells into smaller and more numerous cells is used. While using the same frequencies, the number of transmitter–receivers is increased; consequently the transmission powers are much lower and more subscribers can be serviced. In theory, therefore, the capacity of a cellular system is infinite. In practice, above a certain cell density, the cost per subscriber would become prohibitive and the quality of service would be excessively degraded. Nevertheless, with current analogue systems, cells of 500 m diameter are already used.

The air interfaces of cellular systems are complex and vary from one standard to another (see Chapter 6.3).

On the network side, cellular systems are connected to the public telephone networks, public data networks and other cellular networks as is presently the case in Scandinavian countries and in the future pan-European digital cellular network GSM. The architecture of cellular networks varies from one standard to another. As an example, the general architecture of the GSM network is shown in Fig. 2.6.

The Message Switching Centre (MSC) of the GSM network provides the switching function for incoming calls (from the cable network to the mobiles) and for outgoing calls (from the mobiles to the fixed network). It also manages the data bases of mobiles located within the coverage area of the Base Transceiver Stations (BTS) which are associated with the MSC, subscriptions and the location of subscribers whose nominal base is the MSC.

The Base Station Controller (BSC) provides management of the BTS, concentration of communication between the MSC and BTS thus permitting a reduction of the cost of the links between network elements, supervision of calls and hand-over procedures.

The BTS support radio transmission and reception, control of radio functions and transmission of communications to the BSC. The Operation Maintenance Centre (OMC) provides supervision and control of the

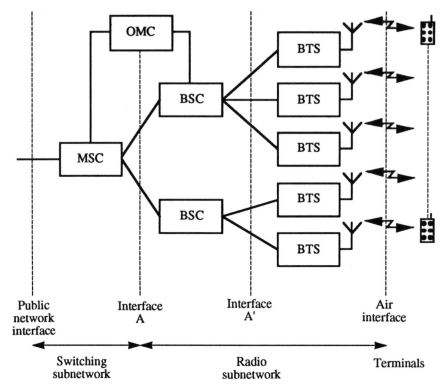

Figure 2.6 General architecture of the GSM pan-European digital system

network elements. It permits configurations and breakdowns to be displayed.

2.4.4 Cordless systems

Technological developments have made the telephone receiver autonomous; it can be identified by a single number. Radio channel access to the network may be via a terminal in the street, a residential base or a base connected to a company telephone system. The various types of cordless system are shown in Fig. 2.7.

This architecture, specific to cordless systems, is thus independent of the analogue or digital standard and frequency bands used.

Increased numbers of telepoint terminals in a public environment and of cordless PABX in a private environment in some cases permit almost continuous coverage to be obtained, as shown in Fig. 2.8, although the low power of cordless terminals limits the range to less than 200 m.

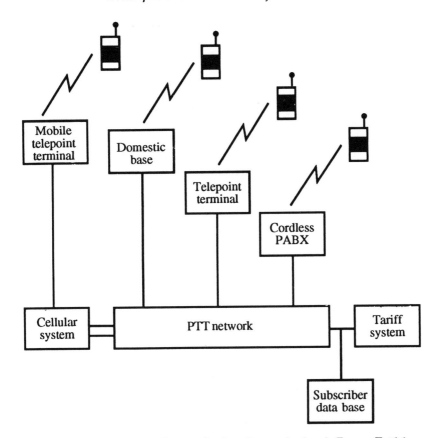

Source : Cordless Communications in Europe. Tuttlebee

Figure 2.7 The various cordless systems

Unlike cellular systems, cordless systems are restricted to radio access to existing cable networks; the cordless standard limits radio interference and defines the communication protocol between the mobile and the access point.

2.5   PRINCIPAL CHARACTERISTICS OF TERMINALS

2.5.1 Terminals

Terminals are described by the two interfaces and their principal characteristics. These two interfaces are the network or system air interface and

Radio coverage zones

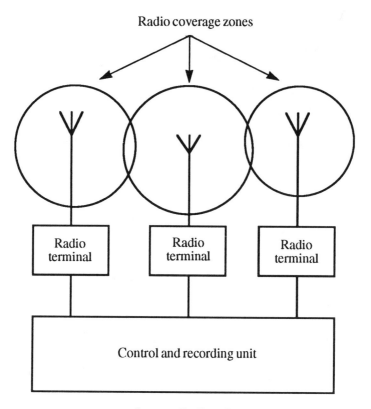

Source : Cordless Communications in Europe. Tuttlebee

Figure 2.8 The coverage of cordless systems

the human–machine interface with the user. The principal characteristics are weight, volume, autonomy, transmission power (which defines the network access coverage) and the functions provided by the associated network such as group calling and call transfer. Autonomy is defined as the number of hours of communication or stand-by for receiving a call or message, without recharging the batteries. Convenience functions associated with the terminal include memory, keyboard illumination, security functions and aesthetics; the last is of increasing importance particularly for terminals offered for sale to the general public, cordless telephones and portable radio telephones.

A quality characteristic related to reliability should be included; this is expressed as an annual rate of return to base of terminals, although this data is generally kept confidential by the manufacturers and remains difficult to measure by an external observer.

2.5.2 Personal paging terminals

The paging terminal is extremely simple in principle. It consists of an integrated antenna, a radio front end which filters the radio channel, amplifies and demodulates the received signal, a logic section, often a single integrated circuit which controls the radio section and interprets the received signal, and finally the human–machine interface which consists of a numeric or alphanumeric screen or bleeper which alerts the user to a call. Fig. 2.9 shows the constituent elements of a pager.

2.5.3 Cellular radio telephone terminals

Cellular terminals are much more complex than pagers. Their principal characteristics are described in Chapter 6.3. They consist of an external antenna for vehicle and portable terminals (sometimes integrated for portable terminals), a filter, called a duplexer, to separate the transmitted and received signals, and a radio demodulation section using an intermediate frequency derived from a frequency synthesiser. For transmission, the signals are modulated and amplified to the power requirements of the terminal; a logic section controls the radio section and the human–machine interface components such as a keyboard and screen. Fig. 2.10 shows the constituent elements of an analogue cellular terminal.

Digital cellular terminals are of greater complexity than analogue ones. This additional complexity is associated with digital signal processing which includes analogue to digital conversion, equalisation (that is, elimination of multiple trajectory effects), deciphering, detection and correction of errors. Furthermore, the services offered on digital networks are much more numerous than on analogue ones with a wide range of supplementary services which enrich the overall human–machine relationship. In practice, the software of a digital cellular terminal is five times greater than that of an analogue terminal.

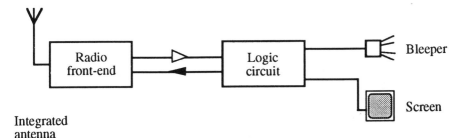

Figure 2.9 The constituent elements of a paging terminal

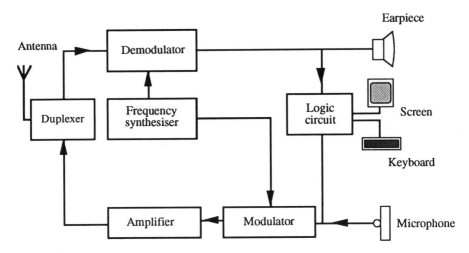

Figure 2.10 Block diagram of an analogue cellular terminal

2.5.4   Cordless terminals

The basic architecture of a cordless terminal is represented in Fig. 2.11. This simplified diagram separates the processing of high frequency signals in the transmitting and receiving channels from the low frequency signals in the audio block.

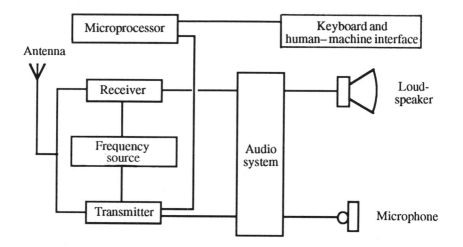

Source : Cordless Communications in Europe. Tuttlebee

Figure 2.11 Descriptive diagram of a cordless terminal

A microprocessor handles all user interface tasks such as the TIM function for receiver and subscriber identification, the dialling keyboard, the transmitting and receiving functions and the communication protocol for establishing a radio channel with a base.

Integration of these products, although already high, is currently limited by the high levels of power consumption required. In fact the size of the integrated circuits and their power consumption requires the use of batteries whose large size is often incompatible with the intended size of the product. More advanced integration of components and batteries in conjunction with assembly technology might permit the size to become that of a pocket telephone.

3 REGULATION AND STANDARDISATION

3.1 INTRODUCTION

The telecommunications sector has become one of the highest priority activities of the European Commission since 1986. Activities in this area are of two types—regulation and standardisation.

The existing national statutory structures, due to the history of telecommunications, have held back most of the political aspects of European projects. This is the reason why one of the first tasks of the commission has been to establish a new set of regulations, which are described in the first part of this chapter.

The single European market assumes the possibility of continuity of service from one country to another together with the disappearance of technical barriers in order to create true conformity of telecommunication products accepted at European level. This objective has necessitated the creation of a European Telecommunications Standards Institute (ETSI), charged with the formulation of European standards, whose new functions are detailed in the second part of this chapter.

The time factor has become crucial in the formulation of European standards. This awareness of a need to act at a European level has become more acute as whole sections of the professional and domestic European electronics industry have disappeared, having succumbed to Japanese competition. Having successfully defeated the United States, Japanese industry turned to the large single market for telecommunications equipment and services created on the first of January 1993.

The stakes are high; telecommunications has become one of the strategic sectors of the European economy influencing all areas of activity. It has been estimated that the gross domestic product of the European Community from the telecommunications sector will increase from 2% in 1984 to more than 7% in the year 2000.

Today this sector represents the direct employment of nearly 1.3 million people in Europe. In the next ten years, 60% of European jobs will depend on the competitiveness of organisations using information technology. The establishment of the single market has, for several years, necessitated the establishment of new regulations and commercial codes of practice. The activity of every telecommunications economy has been disrupted in respect of both equipment, including terminals and system infrastructure, and services.

These elements were the subject of a report published in 1987 by the Commission of the European Community entitled *Green Book on the Development of Common Market Telecommunication Services and Equipment*. This document established a decisive plan of action on a European scale, whose essential objective is 'to develop the conditions under which the market will offer European users a greater variety of telecommunication services of better quality at lower cost'.

The terms of this Community telecommunications policy will permit market competition to be substituted for existing regulations as specified by public authorities or existing monopolies. This deregulation will increase competition for terminals and some telecommunication services, particularly those of radio communication. Access by a greater number of small and medium-sized enterprises to the market will facilitate innovation.

However, terminal equipment remains the difficult point. The following question has still not been resolved: how can the continuity of European industry be guaranteed in a competitive European environment in the face of Asian industries operating on a more organised and less liberal basis?

Regulation and standardisation are the two major sections of the new European telecommunications policy. Substantial national and supernational structures must be developed in order to integrate the single European market and standards must be formulated as the basis for large radio communication projects which permit the creation of a mass market.

These two aspects are the bases required to give Europe a dominant position in the year 2000 in the economic battle for the personal pocket telephone, an electronic product for the professional and the general public, which has not yet been conquered by south east Asian industry.

3.2 STATUTORY ASPECTS

The single European market is characterised by free circulation of people, goods, capital and services. Awareness of the political aspects and strategic importance of telecommunications within the economy led to the publication of the *Green Book* in 1987. This report includes an overview of the telecommunications sector and also a plan of action. A vast debate involving operators, manufacturers and user groups followed this publication. The creation of a single market in 1993 from the present national fragmentation is certainly running against time. Each country has specific traditional regulations and is today confronted with different problems. The use of radio frequencies by public services is one of many examples. The problem of freeing those frequencies which are dedicated to a European mobile radio service must be resolved country by country and case by case. European telecommunications are not uniform and have very large regional and even local disparities. Fig. 3.1 shows several criteria by country which enable these disparities to be appreciated.

The European Commission has presented a four point plan of action for telecommunications:

● 1. The creation of a competitive environment favourable to the development of equipment and services.

● 2. Protection of the role of telecommunications administrations, thereby guaranteeing the continuity of the infrastructure of a basic network.

● 3. Deregulation of bidding for services.

● 4. Promotion of a policy of employee migration from traditional activities to the telecommunications and information technology sector.

These basic objectives are, however, confronted by an increasing complexity of the sector due to the proliferation of communication services. Fig. 3.2 shows the development of telecommunication services between 1847, the year of the birth of the telegraph, and the year 2000. The following five major technological areas have led to the great acceleration of the last twenty years.

● 1. Digitisation of the basic networks in respect of both switching and transmission.

Some European Disparities

Situation at the start of 1993	Germany	Spain	France	Italy	UK
Principal lines/100	48/20	32	48	37	45
Cellular subscribers (thousands)	700	180	450	800	1400
Paging subscribers (thousands)	400	100	300	160	800
Total PMR (thousands)	1200	100	600	400	700
Total cordless (thousands) (1)	1200	600	2200	500	2000

(1) Approved equipment.

Figure 3.1 European disparities

1847	1877	1920	1930	1960	1975	1984	2000
Telegraph	Telegraph	Telegraph	Telegraph	Telegraph	Telegraph	Telegraph	Telegraph
	Telephone	Telephone	Telex	Telex	Telex	Telex	Telex
		Sound	Facsimile	Data	Medium speed data	Packet switched data	Wide band data
			Telephone	Facsimile	Low speed data	High speed data	Packet switched data
			Sound	Telephone	Facsimile	Switched circuit data	Switched circuit data
			Television	High fidelity stereo sound	Telephone	Telemetry	Telemetry
				Colour television	High fidelity stereo sound	Facsimile	Word processor networking
				Mobile telephone	Colour television	Word processor networking	Text facsimile
					Mobile telephone	Videotext	Facsimile
					Personal paging	Telephone	Electronic messaging
						Video conferencing	Newspaper teleprinting
						Sound	Videotext
						High fidelity stereo	Voice facsimile
						Colour television	Telephone
						Mobile telephone	High fidelity telephone
						Personal paging	Teleconferencing
							Video conferencing
							Video telephone
							High fidelity stereo sound
							Quadrophony
							Colour television
							Stereophonic television
							High definition television
							Mobile video telephone
							Mobile telephone
							Mobile text
							Mobile facsimile
							Mobile data
							Mobile Videotext
							Personal paging

Source: Etude CEC.

- 2. The advance of data processing within networks.

- 3. New techniques for information coding.

- 4. Integration of electronic functions into ever smaller miniaturised components of low power consumption.

- 5. The development of new materials and more integrated assembly techniques.

A set of directives has been defined to take account of the existing framework of regulations. Although national structures must evolve towards greater competition, it is fundamental to guarantee the integrity of the basic telephone service. It is therefore necessary to preserve the role of the telecommunications administrations in their provision of an infrastructure to enable them to support the development and maintenance costs of the network. The basic voice telephone service which represents nearly 80% of the turnover of the telecommunication administrations is thus reserved for them. The other telecommunication services are in the competitive domain, particularly radio communication services.

Nevertheless, the statutory and operational functions of current telecommunications administrations must be separated into distinct organisations. On the one hand, the operator will be responsible for the operation of the network, and on the other, a statutory body will be created. These national statutory organisations are described in Section 3.3.4. They have the functions of frequency management, assignment of operating licences, definition of obligatory specifications, checking of agreement procedures and monitoring the observance of conditions of use.

In order to respond to the objectives of the Treaty of Rome, the terminal equipment market must be open to competition. The first condition of marketing equipment for connection to the network remains, however, the homologation procedure or the agreement in force. In order to take account of national disparities, provision of the first wired telephone receiver could temporarily remain exclusively with the operator.

The European market is thus established and directives concerning mutual recognition of the conformity of terminal equipment have been agreed. There are three objectives: to create a single terminal market; to harmonise the procedures for certification, testing, branding and monitoring of products; and finally to guarantee the right to connect legally marketed terminal equipment to public telecommunications networks without additional procedures.

3.3   STANDARDISATION

3.3.1   The importance of standardisation

Standardisation is the stabilising factor of a market economy. The creation
of standards facilitates competition between manufacturers for products
and between operators for services, by guaranteeing to the user that the
product or service purchased conforms to his requirements, has a quality
measured by an independent organisation and will be able to operate with
existing systems. Establishment of standards, or standardisation, is the
often lengthy but indispensable stage of every new technological project.
It must take account of the constraints of all sectors.

There are three levels of the various bodies governing telecommunications standards: international organisations; committees belonging to
the European Community; and national groups belonging to the industry
or telecommunications ministry, the public operator and professional
bodies.

3.3.2   International organisations

The major international organisation is the ITU or International Telecommunications Union. This agency of the United Nations, created in 1866,
today has almost 200 member countries. The ITU is charged with defining
the major activities of the telecommunications sector, not only in the publication of regulations for the telephone, radio and telegraph, but also in
technical recommendations for the definition of standards. However, its
major role is to establish and maintain co-operation and cohesion of the
national policies of the telecommunication administrations of the member
states.

The structure of the ITU presented in Fig. 3.3 shows three temporary
groups: the plenipotentiary conference, administrative conferences and
the administrative council together with four permanent councils; the
CCITT (International Telegraph and Telephone Consultative Committee),
the CCIR (International Radiocommunications Consultative Committee),
the IFRB (International Frequency Registration Board) and the General
Secretariat.

The plenipotentiary conference, which takes place every five years (the
next will be held in 1994), defines the general principles of the organisation
and the directions of its policy. At this time the charter or convention of
the ITU is published. The administrative council convenes annually and
includes 41 members elected by the plenipotentiary council. On the
occasion of the world (CAMTT) and regional administrative conferences,

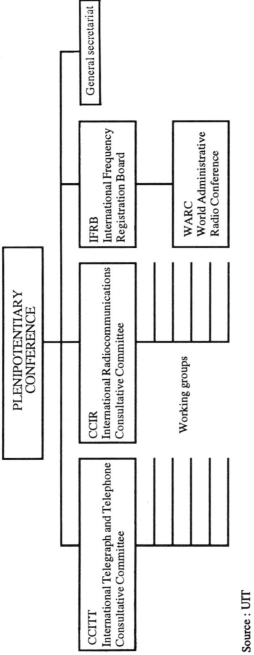

Figure 3.3 The structure of the International Telecommunications Union (ITU)

Source : UIT

practical regulations and the major aspects of work defined in the context of this convention are published.

The CCITT is responsible for international aspects concerning telephone and telegraph networks while the CCIR manages the aspects associated with radiocommunications. These two bodies produce publications, reports and directives. These do not, however, have the force of law at national level but serve as a working basis for European organisations responsible for the establishment of standards. The CCITT and CCIR are managed by full assemblies which convene every two or three years. They also include permanent or semipermanent technical groups responsible for various projects defined by the full assemblies.

The IFRB manages the radio frequency spectrum from 9 kHz to 275 GHz. This organisation is the guardian of international coherence of frequencies through the intermediary of the WARC (World Administrative Radio Conference). WARC manages frequency allocation by dividing the world into three regions: the first includes Europe and the USSR, the Middle East and Africa; the second concerns North and South America together with Greenland; the third region combines the Far East and Oceania. These allocations are of an obligatory nature in these various regions and member states have the responsibility of applying them in accordance with their own timetable. The last WARC conference took place in Spain in 1992 in order to allocate frequency bands for new mobile communication services for the next twenty years.

Fig. 3.4 shows the current distribution of the frequency spectrum between 30 and 3000 MHz.

3.3.3   European organisations

3.3.3.1   CEPT

CEPT, or the European Conference of Postal and Telecommunications Administrations, embraces 26 European countries (EEC and EFTA together with Turkey and Yugoslavia). This organisation was created in 1959, is open only to administrations and has seen its role in telecommunications standardisation partly transferred to ETSI in 1988. The new role of CEPT has been restricted to co-ordination and allocation of radio frequencies.

A committee of CEPT has recently been created, the European Radiocommunication Office or ERO, in order to administer the frequency spectrum on a European plane; the frequency spectrum is an essential resource in the context of the development of new mobile services.

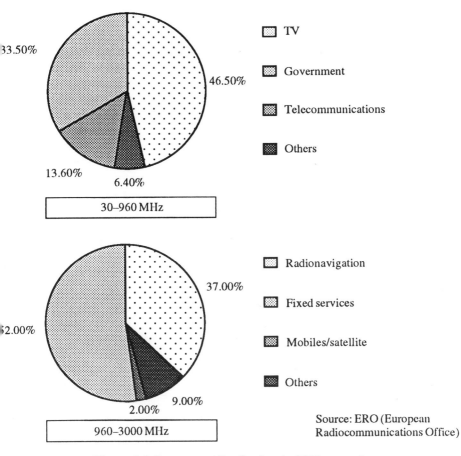

Figure 3.4 Frequency distribution in EEC countries

3.3.3.2   ECTEL

ECTEL, or the European Committee of Telecommunications and Electronic Professional Industries, represents the views of various European national syndicates of the telecommunications and professional electronic industries. This organisation, considered to be the European syndicate of these areas of activity, is a member of ETSI and plays the role of mediator in transmitting requests from European industries via the European Commission.

3.3.3.3   ESPA

ESPA has a role similar to that of ECTEL but is restricted to manufacturers of radio paging terminals.

3.3.3.4   *CEN/CENELEC*

CEN and CENELEC are the European standardisation committees of the electrical and electronic sector. The areas covered by these two sectors and that of telecommunications often make for difficult relations between these various committees, CEPT and ETSI.

3.3.3.5   *ETSI*

The European situation changed in 1987 with the publication of the telecommunications *Green Book*. This report on European telecommunications brought a number of malfunctions to light and established a plan of action to remedy them. The creation of a standardisation organisation is one of the essential factors; ETSI, the European Telecommunications Standards Institute, was born in this way.

In order to accelerate the process of establishing standards and specifications of assent within the community, the compilers of the *Green Book* proposed the creation of a European telecommunications standardisation organisation. This new body, created in February 1988, with the acronym ETSI, has its centre based in France at Sophia Antipolis.

Fig. 3.5 shows the principal direct links between the standardisation committees. It shows the complexity of the standardisation processes. This ponderous structure, due to the history and evolution of technology over

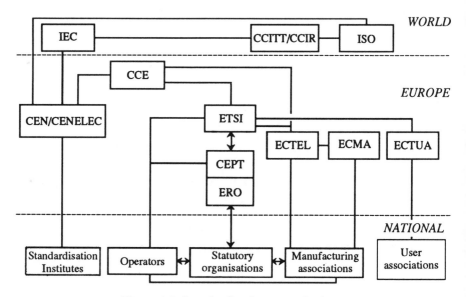

Figure 3.5 Standardisation organisations

a hundred years, has become incompatible with the accelerating progress of technology and the proliferation of telecommunication services.

ETSI is another organisation which makes the existing system even more complex. This independent organisation has been created from the existing structure of CEPT (European Conference of Post and Telecommunications Administration) in order to accelerate the process of telecommunications standardisation. The European market, and every manufacturer, risked being penalised by the delay in the provision of standards. The objective is therefore two-fold: to establish standards for new European products and services in accordance with market requirements and above all to observe the timetable for the establishment of these standards.

ETSI is financed by its members, the number of which exceeded 250 at the end of 1992. Unlike CEPT where only telecommunications administrations can be members, ETSI includes not only these administrations, but also private operators, manufacturers, research centres and user associations. This new structure throws a different light on problems and takes account of engineering, marketing and statutory aspects in the formulation of standards.

ETSI publishes standards of two types: European Telecommunications Standards (ETS) and Interim European Telecommunications Standards (I-ETS). These standardisation plans then become the subject of public enquiries in each of the member countries of ETSI before becoming officially adopted. When an ETS is published, it replaces the corresponding national standard in each member country. When an I-ETS is published, it can co-exist with the corresponding current standard in each country.

ETSI has a general assembly which has the essential role of defining the policy of ETSI and determining the operating budgets by vote in the light of the programme of work adopted by the technical assembly. It is this technical assembly which prepares the annual study programme and adopts the ETS and I-ETS after public enquiries. Its decisions are determined as the weighted majority of national colleges, following a mechanism inspired by the operation of the EEC. The national weightings used during voting have been established from the telephone inventory of each country.

The weightings of votes of the European Community (EEC) and the Free Trade Area (EFTA) are: France (10), Germany (10), Italy (10), United Kingdom (10), Spain (8), Belgium (5), Greece (5), Netherlands (5), Portugal (5), Denmark (3), Ireland (3), Luxembourg (2), Sweden (5), Switzerland (5), Austria (3), Finland (3) and Norway (3).

The standardisation work is performed in technical committees. These meet several times per year and consist of co-opted experts, seconded for a limited period by member companies, and specialists working part time for ETSI. Their work leads to proposals for standards which are subjected to public enquiry after approval by the technical committee.

3.3.4   National organisations

There are two types of national organisation: the body in charge of agreed procedures whose statute varies from one country to another, and syndicates of telecommunications manufacturers. Some significant examples will be given here.

Each country has its manufacturers' syndicate or syndicates, according to the area of activity, which are all members of the European syndicate ECTEL. France, for example, has two, SIT (Telecommunications Industries Syndicate) and SPER (Syndicate of the Professional Electronic and Radio Equipment Industry). Historically, the telephone sector occupied by SIT was very different from the radiotelephone sector, of which SPER is in charge. The evolution of mobile services and recent developments in cordless telephones should lead to a new deal for this syndicate.

Today, national bodies in charge of the regulation of telecommunications exist in Europe and have differing status. The reason is not deregulation during recent years under the influence of the European Commission. The prior condition relating to competition, made clear in the *Green Book* of 1987, of separating statutory activities and those of the operator, has been progressively put in place. The cases of OFTEL in the United Kingdom and DRG in France will be cited here.

OFTEL   The United Kingdom Office of Telecommunications, OFTEL is an independent organisation created by a law of 1984 (*The Act*). It is financed by Parliament and the taxes received from issued licences. The key functions of OFTEL are to ensure that licence holders respect their initial agreements and to define the procedures relating to the approval of telecommunications terminals.

DRG   The French organisation is DRG, or Directorate of General Regulations; it was created in 1988 and depends directly on the Ministry of Telecommunications. It was not until the beginning of 1991, with the new status of France Telecom, that DRG has been effectively separated from France Telecom within the Ministry of P & T. DRG has three major roles: provision of authorisation to operate networks and telecommunication services; approval of terminals; and approval of procedures for admission of installers.

3.4   STANDARDISATION OF RADIO COMMUNICATION

3.4.1   The Strategic Review Committee

New systems of radio communication are one of the priorities clearly addressed by the technical assembly of ETSI. Each year this assembly chooses one aspect from the major projects in progress for particular consideration. A complete assessment must then be produced, in several months, by a committee called the Strategic Review Committee (SRC). The first actions of ETSI have related principally to ISDN between 1986 and 1990. Standardisation effort was superseded from 1990 by terrestrial and satellite radiocommunication services with mobiles. The SRC report of 1991 concerning the evolution of new services identifies two groups, in the short and medium term respectively, according to the date of commercial opening. The first group includes GSM projects for cellular, DECT for the cordless telephone and ERMES for one-way radio messaging. The second group of services is more speculative and is under discussion; it seeks to respond to the overall requirements of mobile communications. PCN is one part of the response.

The SRC determines the development of mobile services by division into two aspects as presented in Fig. 3.6. The first aspect separates mobile services aimed at the mass market from specific services for niche markets while the second aspect distinguishes existing services from new ones.

The SRC report has made evident a multiplicity of new services and the need to create new technical working groups specifically for sufficiently mature projects. One of the conclusions remains the difficulty of regulating different generations of services which often cover the same target markets. ETSI must thus ensure standardisation of GSM, PCN and DCS

	Established markets	New markets
Mass markets	Cellular Cordless telephone Public radio paging Private radio paging Radio paging with confirmation	Telepoint Cordless PABX Traffic information Personal telephone
Niche markets	Private radio (PMR) Citizens' band (CB) Satellite radio paging Data transfer by satellite Telecommand and telemetry Maritime systems	Data transfer Trunk PMR Service provider PMR DSRR Satellite cellular systems (LMSS) Satellite trunked PMR Aeronautic systems

Source: ETSI/SRC/Mobile Expert Group 1990.

Figure 3.6 Division of ETSI services and markets

1800, DECT, ERMES and UMTS so that these projects can have complementary objectives.

3.4.2 ETSI projects

3.4.2.1 *The project structure*

Four technical committees cover all new radio communication services and equipment. These ETSI groups, whose structure is given in Fig. 3.7, are RES, GSM, PS and SES.

3.4.2.2 *Radio Equipment and Systems (RES)*

This technical committee is the oldest and that in which all ventures have started. New and dedicated committees have been created as a consequence of the advance and importance of the various projects. RES is currently divided into six activities:

● RES 1 is concerned with maritime radio equipment including the following products.

 1. The VHF radio telephone.
 2. Direction finding and distress markers at 406 MHz.
 3. Portable VHF direction finding and distress beacons.
 4. Radar at 9 GHz for rescue operations.

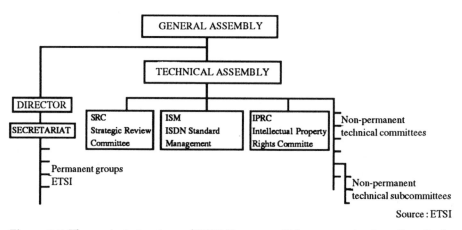

Source : ETSI

Figure 3.7 The project structure of ETSI (European Telecommunications Standards Institute)

- RES 2 is concerned with terrestrial mobile radio hardware. The subjects currently under examination and which are the subject of public enquiries are as follows.

  1. Analogue equipment with a removable antenna.
  2. Selective call signalling.
  3. Data transmission equipment with integral antenna.
  4. Measurement techniques for radio parameters and determination of error boundaries.

- RES 3 oversees short range mobile radio equipment as follows.

  1. The Digital Short Range Radio (DSRR) system without a fixed infrastructure at 900 MHz.
  2. The Digital European Cordless Telecommunication (DECT) system for residential, telepoint and PABX applications.

- The activities of RES 4 have been taken over by the Paging System (PS) technical committee.

- RES 5 is responsible for radio telephone systems which can be used by passengers in airliners by means of a terrestrial infrastructure such as the Terrestrial Flight Telephone System (TFTS).

- RES 6 is responsible for shared digital resource systems (trunks).

3.4.2.3 Group Special Mobile (GSM)

This group formulates the European digital radio telephone network standard GSM at 900 MHz. The enormous amount of standardisation work has been separated into five subgroups charged with service aspects, radio interfaces, network aspects, data transfer and definition of the Subscriber Identification Module (SIM), a memory card which contains the data on the subscriber account, and the service. The very short time scale for standardisation of this project requires the support of a permanent 'project team 12'.

3.4.2.4 Paging System (PS)

This group is responsible for standardisation of the new European wide area paging system called ERMES.

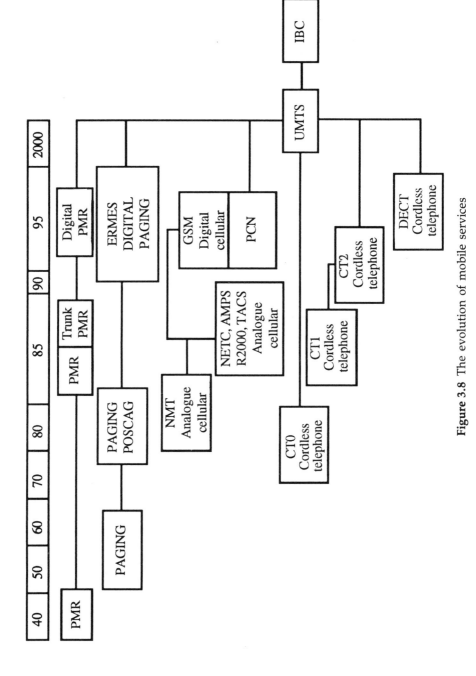

Figure 3.8 The evolution of mobile services

3.4.2.5 Satellite Earth Stations (SES)

This fourth committee standardises ground stations for satellite reception.

3.4.2.6 Developments

The report of the Strategic Review Committee published at the beginning of 1990 has brought certain critical points to light. Twenty-seven recommendations have been presented to the technical assembly and six technical subcommittees have been created which depend on the RES group to put these recommendations into practice.

- 1. RES 8 (Low Power Devices).

- 2. RES 9 (EMC) to examine the problems of electromagnetic compatibility.

- 3. RES 10 to examine local radio networks, called HYPERLAN for interconnection of computers by means of radio links.

- 4. RES Ad hoc UMTS (Universal Mobile Telephone Service) to anticipate standardisation of the universal mobile telephone.

- 5. RES Ad hoc Spectrum Needs, to examine subsequent requirements of the frequency band used for new radio services.

- 6. RES Ad hoc CT2 telepoint to be responsible for standardisation of the CT2 programme described in Chapter 7.4.

- 7. RES Ad hoc Radio Site Engineering.

3.4.3 Standards and authorisation

The key point of Europe in 1992 remains the principle of European standards based on mutual recognition of the authorisation of terminals. A new organisation has thus been created, ACTE, or Approval Committee for Terminal Equipment; it is responsible for certification of European approval and also quality assurance procedures for manufactured products and the associated manufacturing processes. Fig. 3.8 shows the evolution of mobile services and the associated standards in the three areas of paging, the cordless telephone and the cellular radio telephone.

Part 2

MOBILE SERVICES AND PRODUCTS

4 RADIO PAGING

4.1 INTRODUCTION

4.1.1 The concept

The paging situation varies greatly throughout Europe. Many systems based on different technical standards have been installed with varying degrees of commercial success; penetration rates can vary by a factor of ten between European countries.

The existing systems are often incompatible and do not permit economy of scale at infrastructure or terminal level. Furthermore, product and service prices, which depend on operators and distribution, lead to very different market sizes. While Great Britain has more than 800 000 subscribers, that is a penetration rate of almost 15 per 1000 of the overall population, other countries such as Spain and Italy have not reached the 3 per 1000 barrier. Some operators are suggesting an international subscription from the start of 1991, but the truly European radio message system, which uses ERMES digital technology, will not be operational until 1993.

In the first part of this chapter, existing technical systems and their geographical locations will be described. The principal European markets will then be examined by analysing available products and services, and market demand in 1993. The second part will examine the position of Europe in relation to the rest of the world, particularly the United States, Japan and other Asian countries together with the ERMES project; this is managed under the sponsorship of the European Commission and is a better prospect for paging in comparison with other mobile services.

4.1.2 On-site paging

Private or on-site paging permits communication on a site limited to a company or site such as a hospital, an airport, a factory or offices, for internal use; it requires the prior granting of an operating licence by the national organisation responsible for frequencies (DRG, OFTEL). On-site paging services are either unidirectional with messages sent to a mobile receiver or bidirectional which makes voice communication possible. In 1993 these remain the only optional mobile services available for existing private automatic branch exchanges (PABX) for persons moving within an organisation.

It will be seen in Chapter 7.3 that new versions of the PABX, with the introduction of cordless technology, will have a very strong influence on the development of this specific market.

4.2 PUBLIC PAGING

4.2.1 The principle

Among the numerous public mobile radio communication services, there are personal paging systems or one-way radio message systems which, up to 1989, had the greatest number of subscribers throughout the world. The number of subscribers is currently estimated at almost 30 million in the world, of which 15 million are in the United States and 3 million in Europe.

The one-way radio pager has been ousted from its position as leader of mobile services by the exceptional increase of cellular systems. The number of subscribers to cellular systems in Europe is of the order of 6 million compared with 3 million radio paging subscribers.

The advantage of paging over other radio communication services lies in the simplicity of its concept; messages are sent via a radio channel to a receiving terminal situated within a given area. The receiver is very compact and can be slipped into a pocket, a bag or hooked to a belt and in some cases even combined with a watch or pen.

Each terminal is uniquely identified by a code. A correspondent, knowing the code, can at any time request to send a message to the receiver via a telephone or videotext terminal. This message is routed by the switched telephone network or an X.25 network to a central server and then transmitted on a radio channel into a given region. The area over which radio transmission is achieved depends on the system used and the subscription contract. Some systems are national, that is the message is transmitted over the whole of the territory while others limit transmission to one region; the territory is thus divided into different zones. The correspondent must thus have prior knowledge that the subscriber is within the subscription area to be certain that the message will reach him.

Communications using a public paging service require an advance rental subscription together with the purchase or hire of a receiver. A public subscription is obtained either directly from the operator or through an authorised distribution network.

The service operator has an operating licence for the use of one frequency band which is limited to a given geographical area, regional or national.

4.2.2 Different types of message

Three types of message can be received in accordance with the system used and the terms of the subscription: a beep, a numerical message or an alphanumeric message.

In the first case (a beep), the service is limited to sending basic signals. The receiver reacts by emitting a tone signal (a beep), illuminating a diode or vibrating in the case of more sophisticated apparatus. The user knows that someone wishes to contact him, but he cannot determine the origin, the urgency or the reason for the call. Generally, users subscribing to such a service give their code to only a few correspondents such as a secretary, a partner or an associate. It is then possible to determine the origin of a call and to call the correspondent back.

The level 2 service (numeric) permits reception of exclusively numeric messages. The correspondent sends a number using the keyboard on his telephone. This service is generally used by sending either a telephone number to be called or a number associated with a code determined in advance between the correspondent and the subscriber, for example, 1, 2 or 3 for calls of different degrees of urgency.

The level 3 service uses transmission of alphanumeric messages consisting of a text of several words which is displayed on the receiver screen. The text is composed by means of a videotext terminal (Minitel in France, Bildschirmtext in Germany, Prestel in the United Kingdom) or a microcomputer connected to the telephone network.

The more sophisticated receivers are programmed in advance to react in a specific manner when they receive a given message. They display preprogrammed icons, from a library, in accordance with a received code. The received message can thus be text combined with a graphic image of a purse, a telephone, a restaurant or a company logo.

4.2.3 A contrasting situation

The advent of personnel locating services remains a partial success. Although the number of subscribers to public paging services throughout

Europe will increase between 1993 and the year 2000 from 3 million to more than 10 million, the increase in this sector for the coming years will remain less than that of the radio telephone or the cordless telephone described in Chapters 6 and 7.

The price factor is not directly the cause since the cost of using a paging service, including amortisation of the receivers and the monthly subscription to the operator, is the lowest of all the radio communication services (see Fig. 4.1).

This one-way service can be considered in some cases to be half a service. Although it responds perfectly to the specific requirements of urgent calls for professionals subject to compelling obligations, such as the medical professions and technical security services, its use often remains frustrating in comparison with two-way telephone communication for the majority of persons on the move. The use of a pager is effectively passive. The subscriber, in a stand-by situation, can receive messages but cannot respond instantaneously.

Furthermore, if one compares not simply the mean cost of use but the mean cost of use in relation to the volume of data received (numeric, alphanumeric or voice), that is the service provided, it is evident that the paging service is dearer than the radio telephone service. A simulated comparison is presented in Fig. 4.2.

4.2.4 Developments

The personal paging service will undergo several important developments in the coming years. This change will arise from technology and also the maturity of the sector. It will be possible to integrate the personal paging

Type of service	Equipment purchase	Subscription/ month	Cost of connection	Call cost/ month	Annual cost
Cable telephone	£60 inc. tax	£1.50 inc. tax	£60 inc. tax	£25	£350
Radio paging	£300 inc. tax	£20 inc. tax	£30 inc. tax	£0	£350
Cellular	£1200 inc. tax	£30 inc. tax	£65 inc. tax	£120	£2000
Telepoint	£200 inc. tax	£5 inc. tax	£30 inc. tax	£50	£750

The costs in this table are average values for the various countries of Europe and give only a general indication of the services with respect to each other.
This comparison is based on average market levels for the equipment and services.
Purchase of the equipment is amortised over three years.
Radio paging is assumed to be alphanumeric.
The bonus taken by the service providers in the United Kingdom is not taken into account.

Figure 4.1 Comparison of the costs of mobile services

Type of service	Monthly cost	Calls/month	Words/call	Cost of transmitted word
Radio paging	£30	120	5	5 p
Cellular	£170	120	250	0.5 p
Telepoint	£60	120	250	0.2 p

The mean monthly cost is obtained from the table of Figure 4.1.

Figure 4.2 Comparison of the costs of mobile services as a function of the messages transmitted

service into radio communication equipment as an option or as an additional function.

Evolution of demand can be seen at the present time. Although videotext terminals are not as widespread in all countries of the Community as they are in France, the demand for alphanumeric services is increasing gradually so that it exceeds that for basic services (75% in France, 50% in Spain).

Personal paging is increasingly offered with additional services such as voice messaging. The subscription then includes the telephone number of a 'vocal letter box' where all correspondents can leave a message. This voice messaging service complements radio paging. The receiver can also indicate to the subscriber that a message has been deposited in his letter box; a personal code number then permits him to listen at any time from a conventional telephone.

The major development will arise from other mobile radio services. The radio paging receiver will be integrated into radio telephones or new urban cordless telephones of the telepoint type as described in Chapter 7. Paging will be considered as an option in the same way as an answering machine or hands-free operation. In this case, locating persons on the move is one of the elements of a more complete commercial proposal.

4.3 TECHNICAL SYSTEMS

4.3.1 Eurosignal

The Eurosignal service operates in accordance with the same technical specification in Germany, France and Switzerland. In service since 1974, it now has more than 300 000 subscribers in these three countries and the subscription may be either national or international.

The German and French systems are divided into regions. One transmission frequency, and hence one radio channel, corresponds to one region. It is thus necessary to tune the receiver to the radio channel corresponding to the region visited. Although Eurosignal coverage is international, in order to call a subscriber it is in every case necessary to know which region he is in.

The coding principle used by Eurosignal limits the service to the reception of 'bleeps' but it is nevertheless possible to identify several correspondents by means of different subscriber numbers and an indicator on the equipment (light emitting diodes or a display). The coding used for Eurosignal was defined in 1970 by the German telecommunications administration and uses amplitude modulation at 87 MHz.

4.3.2   POCSAG

The Post Office Code Standardisation Advisory Group (POCSAG) standard was drawn up in 1978 by British Telecom and various manufacturers. This group, called POCSAG, was formed on the initiative of British Telecom and pragmatic British policy to unite participating pager manufacturers and to provide the opportunity of creating a new standard in response to the requirements of these new services. However, British Telecom has allowed third party manufacturers of POCSAG the freedom to use this new standard.

This code has not been protected by patents so that it can be widely used and become an effective international standard. In 1982, the world radio communications standardisation organisation, the CCIR, recognised it as the first standardised international code for pagers: RCP1 or CCIR Radiopaging Code No 1. In 1983, then in 1984, the POCSAG coding system was enhanced with numerical and alphanumeric message services respectively. RCP1 or POCSAG is today the system most widely used throughout the world.

4.3.3   GOLAY

Although the POCSAG code enjoys access free of all patents, there are variants developed by some organisations which optimise particular aspects. These coding modifications are for technical reasons but they also serve to evade free use of the POCSAG patents. These codes, derived from POCSAG, effectively become specific systems which are not compatible with other systems. Their major use is in the creation of a captive terminal market once an infrastructure has been installed. The company responsible for installation of the system thus controls the terminal market.

One example of a specific code is the GOLAY code developed by Motorola whose exclusive property it remains. Widely used in the United States and Australia, it is also used in Italy, Spain and Austria.

4.3.4 Radio Data System

The RDS system has made use of an original principle. It uses the sub-carrier of the FM radio signals transmitted on existing national networks. It is currently used in only a few countries such as Sweden and France. Simple in principle, its major advantage lies in the technique used to combine two complementary services, radio messaging and radio broadcasting. With an FM broadcasting network covering all of the national territory, it is merely necessary to combine the messages to be sent with the sound radio transmissions. The only constraint is that the FM radio transmissions must not be disturbed.

This system can be installed very rapidly and at low cost if an FM radio infrastructure exists. This advantage permits the operator to achieve profitable operation in two financial years with only a few tens of thousands of subscribers. RDS thus perfectly suits developing countries wishing to install a radiopaging system in urban areas. RDS supports text messages in addition to paging.

However, the weak point of the RDS system lies in the high cost of terminals. This high cost is not due to the technical complexity of the products but to limited broadcasting of the RDS system compared with POCSAG; hence the quantities produced by Nokia, the only manufacturer in a position of near monopoly, are small.

4.4 NATIONAL MARKETS

4.4.1 Germany

Two one-way radio message services are currently in service, both a monopoly of Telekom: Eurosignal and Cityruf.

The older service, Eurosignal, was put into service in 1974. Its subscribers can extend to France and Switzerland which use a comparable system. Although Eurosignal is limited to sending 'beep' messages, the German system permits differentiation between the calls of eight different correspondents. In fact the receiver can contain up to eight diodes, each corresponding to a particular radio channel.

The Cityruf service conforms to the POCSAG standard. This second service was launched in March 1989 by the Bundespost to create a new service with transmission of alphanumeric messages. The objective is eventually to replace the ageing Eurosignal system with the higher performance POCSAG system and provide more integrated and less expensive terminals. However, the market only rarely reacts instantaneously and the two services are still experiencing comparable increases.

Competition between these two services at present gives the advantage to Eurosignal for the rate of coverage and to Cityruf for the price of terminals and running costs, but the progressive spread of Cityruf coverage will soon make the last advantage of Eurosignal disappear. Fig. 4.3 shows the essential characteristics of these two services.

Present manufacturers of terminals are mainly Multitone and Philips for Eurosignal; and Motorola, NEC and Swissphone for Cityruf. The receivers are presently 30% distributed by Telekom agencies under its own name while the professional networks distribute all brands.

4.4.2 Spain

One-way radio message operators are very numerous in Spain, of the order of sixty for 90 000 subscribers, but only Telefónica is authorised to provide a service with national coverage. Its new service Mensatel, opened in 1989, has 50 000 subscribers. It uses an NEC POCSAG system on a frequency of 149 MHz with 45% of subscriptions for 'beep', 35% alphanumeric and 20% numeric. NEC also occupies the POCSAG terminal market with an 80% share.

Among the multiplicity of operators limited to local or regional coverage, the main ones are Mensafonico, Telemensaje, Avisa, Busc-person and Aircall for a total of 40 000 subscribers. Mensafonico offers an original voice

Germany	Eurosignal	Cityruf (1)
Year of origin	1974	1989
No of subscribers (1993)	200 000	200 000
Messages	Bleep	Alphanumeric
Basic service	National	Regional
Mean receiver price (inc. tax)	£450	£250
Monthly subscription (inc. tax)	£7	£15
Mean annual cost (inc. tax) (2)	£250	£280

(1) Cityruf cost calculations were made with alphanumeric messages and the basic regional service.
(2) The annual cost includes amortisation of the receiver and connection costs over three years.

Figure 4.3 Radio paging in Germany

Spain	Mensatel (1)	Telemensaje
Year of origin	1989	1986
No of subscribers (1993)	55 000	25 000
Messages	Alphanumeric	Alphanumeric
Basic service	Regional	Local
Mean receiver price (inc. tax)	£250	£250
Monthly subscription (inc. tax)	£40	£40
Mean annual cost (inc. tax) (2)	£600	£600

(1) Mensatel cost calculations were made with alphanumeric messages and the basic regional service.
(2) The annual cost includes amortisation of the receiver and connection costs over three years.

Figure 4.4 Radio paging in Spain

message service with a maximum duration of 15 s with dedicated terminals manufactured by Multitone and Motorola. Telemensaje covers the Madrid region with a Motorola system and offers its 18 000 subscribers the three levels of service.

This fragmented network situation will not evolve appreciably before 1994 and the installation of the ERMES European system. It can be noted, however, that the major manufacturers of this sector such as NEC and Motorola have already taken a dominant position in Spain without a single local organisation having been able to establish any manufacturing activity.

4.4.3 France

The Eurosignal service was launched in 1975 by the PTT at approximately the same time as in other European countries. In 1987, France had 70 000 subscribers while the United Kingdom had 400 000 and West Germany 150 000. This very clear delay caused the Telecommunications minister to increase the availability of services. The French statutory structure consequently suffered a major disruption in 1987 with the authorisation of the opening of 'paging' services to competition. Two services have consequently been opened, the operating licences having been granted to France Telecom under the name Alphapage and to TDF under the name Operator. At the start of 1993, France had nearly 300 000 subscribers for the three services of which 35% were Eurosignal, 45% Alphapage and 15% Operator.

The repurchase of TDF by France Telecom in 1988 favoured a complementary position of these services in their quoted price, service coverage zones, types of message and subscription, although they remained competitive for the end user. Fig. 4.5 shows the essential characteristics of these three services available in France.

The establishment of concerted competition has extended the services offered, but although the French penetration rate is equivalènt to that of Germany, it remains a third of that of the United Kingdom and a quarter of that of the Scandinavian countries.

The Eurosignal service, whose major advantage remains its exceptional coverage, is out of date today. It supports little more than a basic service (beep) and the price of its receivers, using outmoded technology, are the highest on the market. The total service of Eurosignal is stagnating. The commercial competition between offers to new subscribers is therefore concentrated essentially on Alphapage and Operator.

In practice, the competition between operators is not so direct. Their marketing positions are slightly different and a client's choice tends towards one or the other according to his needs. The essential difference between Alphapage and Operator lies in the level of coverage of the service; it is of the regional type with the possibility of transfer between different zones for Alphapage, while Operator is a service with national coverage from the start. The distribution of subscribers is very unfavourable to Operator. This imbalance is explained by the distribution strength of France Telecom offering the Alphapage service. This difference could disappear in the coming years with the integration of Operator into France Telecom.

Another competitive factor in favour of the France Telecom service with respect to that of TDF is the price of a terminal. Although the global installation cost of the RDS system (Operator) is low, since message transmission uses an existing FM network, the volume effect has not yet worked for RDS receivers as it has worldwide for terminals of the POCSAG

France	Eurosignal	Alphapage (1)	Operator
Year of origin	1975	1987	1987
No of subscribers (1993)	115 000	160 000	50 000
Messages	Bleep	Alphanumeric	Alphanumeric
Basic service	National	Regional	National
Mean receiver price (inc. tax)	£450	£250	£450
Monthly subscription (inc. tax)	£7.50	£8.50	£35
Mean annual cost (inc. tax)	£250	£190	£550

(1) Alphapage and Operator costs were calculated for alphanumeric messages and the basic regional service in the case of Alphapage.
(2) The annual cost includes amortisation of the receiver and connection costs over three years.

Figure 4.5 Radio paging in France

standard. Whereas sales of RDS products are several tens of thousands of units in France and Sweden, POCSAG terminals are sold in several millions in numerous European countries. RDS terminals remain at a price 50% higher than for POCSAG terminals. This difference is also explained by the structure of the market; there is only one manufacturer, Nokia, in a monopoly position, of RDS, while POCSAG is subject to cut-throat competition between numerous European, American and Japanese manufacturers.

It should be noted that the Finnish manufacturer Nokia provides the system and terminals for the TDF Operator system while the American company Motorola supplies the Alphapage system to France Telecom. In the more competitive Alphapage terminal market there is not only Motorola with 30% of the market but also Alcatel with 40% and Swissphone with 20%.

4.4.4 Italy

In 1993 SIP maintains the monopoly of radio paging services in Italy. The number of subscribers to the three services (Vehicular, Teledrin and Euromessage) is of the order of 150 000.

The Vehicular service was introduced in 1972. It permits only 'beep' messages to be sent, but offers the possibility of distinguishing the origin of four different correspondents by means of an addressing system similar to that of Eurosignal. Its coverage is almost national using a frequency of 160 MHz.

The truly commercial service is Teledrin with almost 140 000 subscribers compared with 1000 subscriptions for Vehicular. The Teledrin Service uses

Italy	Teledrin (1)
Year of origin	1984
No of subscribers (1993)	140 000
Messages	Numeric (3)
Basic service	Regional
Mean receiver price (inc. tax)	£250
Monthly subscription (inc. tax)	£7
Mean annual cost (inc. tax) (2)	£170

(1) Teledrin costs were calculated for numeric messages and the basic regional service.
(2) The annual cost includes amortisation of the receiver and connection costs over three years.
(3) Alphanumeric service with Euromessage.

Figure 4.6 Radio paging in Italy

the GOLAY system developed by Motorola and derived from the POCSAG system. Subscribers are divided into 70% for numeric messages and 30% for 'beep' messages.

The GOLAY system operates only with specific terminals developed by Motorola. However, to encourage competition in the terminal market demanded by the operator, the GOLAY system delivered to SIP has been modified to permit the use of POCSAG terminals, but only for 'beep' messages. This minor development should not disrupt the commercial semi-monopoly which Motorola exerts on the Italian market.

The latest development concerns the opening of the European Europage service in 1991 which offers the three levels of service, but has only a few thousand subscribers.

4.4.5 The United Kingdom

The United Kingdom has the highest rate of penetration of radio message services in Europe. The number of subscribers is of the order of 800 000, equivalent to the combined totals of Germany, France and Spain.

Several national operators using the POCSAG standard are in competition. The following are the notable operators: British Telecom (BTMC), Aircall/DMC, Mercury, Vodapage and Hutchison.

BTMC (British Telecom Mobile Communications), the subsidiary of British Telecom responsible for paging, supports nearly 500 000 subscribers. As saturation of its network approaches, the commercial policy of BTMC is now tending more towards an improvement of quality for its subscribers (alphanumeric subscriptions, high power, reduced rate of return of terminals) than to a pure increase.

Aircall, following its merger with DMC Paging in 1991, has reached almost 150 000 subscribers. This operation has pushed Air Call, a subsidiary of Bell South, one of the American 'Baby Bells' resulting from the dismantling of AT & T, to the second national rank after British Telecom.

The Mercury subsidiary, Mercury Paging, has 60 000 subscribers and offers its customers a radio message receiver to provide a service of information flashes, 'Messenger Services', and financial information from the *Financial Times*.

The two other operators are Racal Vodapage, a Racal group subsidiary with 80 000 subscribers and national coverage, and Hutchison, which operates a service in the London region and the North of England with 15 000 subscribers.

The European Europage service is operated by all local operators with the exception of Mercury which prefers to await the ERMES system.

Although the United Kingdom has the largest total in Europe, the national manufacturers have not profited from this considerable market.

United Kingdom	BTMC	Aircall/DMC	Vodapage	Mercury Paging
Year of origin	1983	1987	1987	1987
No. of subscribers (1993)	450 000	150 000	80 000	60 000
Messages	Alphanumeric	Alphanumeric	Alphanumeric	Alphanumeric
Basic service	Regional	Regional	Regional	Regional
Mean receiver price (inc. tax)	£250	£250	£250	£250
Monthly subscription (inc. tax)	£20	£20	£20	£20
Mean annual cost (inc. tax) (2)	£300	£300	£300	£300

(1) Cost calculations were made for alphanumeric messages and the basic regional service.
(2) The annual cost includes amortisation of the receiver and connection costs over three years.

Figure 4.7 Radio paging in the United Kingdom

The manufacture of radio paging terminals is very concentrated. Among the five foremost suppliers of the English market (NEC, Motorola, Philips, Multitone and Panasonic), the Japanese company NEC is the leader with 70% of the market and the American company Motorola has 20%. This supremacy covers the whole range of products including 'beep', numeric and alphanumeric terminals.

The tendency will only be to accentuate this supremacy since the large economies of scale achieved by volume production give NEC, and to a lesser extent Motorola, a strong competitive advantage in overcoming costs. For this reason NEC can determine the United Kingdom market price, a burden to which its competitors must submit.

4.4.6 Other European countries

Figure 4.8 shows the European radio paging situation in 1991. It shows a diversity of penetration rates from 20 per 1000 in the Netherlands to less than 1 per 1000 in Portugal and the significance of the STAR (Special Telecommunications Action for Regional Development) programme which has the goal of encouraging development of telecommunications services in disadvantaged regions of the European Community.

Situation at the start of 1993	Subscribers (thousands)	Penetration rate (%)
Austria	100	1.3
Belgium	180	1.8
Denmark	60	1.2
Finland	45	0.9
France	300	0.5
Germany	400	0.5
Iceland	4	1.6
Ireland	12	0.3
Italy	160	0.3
Luxembourg	7	1.8
Netherlands	350	2.3
Norway	110	2.6
Portugal	30	0.3
Spain	100	0.3
Sweden	130	1.5
Switzerland	50	0.7
United Kingdom	800	1.4

Figure 4.8 Radio paging in Europe

4.5 DEVELOPMENTS

4.5.1 European paging

The majority of current developments aim to create a truly European personal paging system. The reasons are simple.

Users are looking for ever smaller receivers of lower cost. This integration arises from the development of specific components to reduce the effective volume, power consumption and price. The new products use only one dedicated component and are much more similar to consumer electronic hardware, such as calculators and watches, than the professional products which existed several years ago. The price can thus be significantly reduced provided that they cover a larger global market. The profitability of such developments follows from the establishment of multinational standards. Numerous projects are in progress, both short term, with international agreement already signed with existing standards (the Europage/POCSAG project), and medium term, with the establishment of new standards recognised on a European scale (the ERMES project).

4.5.2 Europage

To be able to offer a European radio paging service very rapidly at minimum cost and without substantial industrial and technological risk, some operators have moved towards the Europage project which uses the existing POCSAG standard which is the most widespread in the world.

This is why, in 1988, Italian, German, French and British operators signed an agreement to open a radio message service using the POCSAG standard on the common frequency of 466 MHz. This agreement resulted in the commercial opening of the Europage service in 1991; this permits messages to be received in the four signatory countries with a single subscription and a single terminal. This service is based on offering POCSAG systems on services such as Bundespost's Cityruf, France Telecom's Alphapage, SIP's Teledrin and Europage which belongs to a British consortium combining all the existing British operators with the exception of Mercury.

This European analogue radio paging project may seem risky in the face of the competing digital project ERMES. However it clearly reflects the operators' dilemma in view of the technological distribution of the new digital approaches. Mercury has deliberately chosen to await the arrival of the ERMES project, public operators such as France Telecom and Telekom are preparing to manage the transition with two competing offers and several private operators are investing only in Europage, a short term solution and a development of their analogue networks.

4.5.3 ERMES

To avoid creating incompatible national radio paging networks, ETSI, under the aegis of the European Commission, has specified and planned the introduction of the ERMES project which is the first truly European radio paging system. The objective of the Commission is two-fold: to create concerted competition for terminals and services. The setting up of a single standard which can operate on all European networks must very rapidly give manufacturers economies of scale, lower prices and thus very competitive products with European approval for consumers. ERMES also permits competition for services since this system permits a multiplicity of operators in the same country. Hence the customer will be able to choose an approved receiver from several manufacturers and a service from several operators by comparing the prices and benefits provided.

In 1990, sixteen operators in eight European countries signed a formal agreement relating to the timetable for setting up the ERMES system. At the same time, the European Commission was preparing the directives for installing the system, notably the freeing of frequencies. The frequency band at 169 MHz is to be reserved for the service and the timetable for development with respect to the various EEC countries is as follows:

- December 1992: opening of the service.

- January 1994: the system must cover a minimum of 25% of the national population.

- January 1995: minimum coverage of 50%.

- January 1997: minimum coverage of 80%.

Although this project is well advanced, technical delays in system development and statutory ones in establishing European policy have arisen. Market studies, however, predict a strong market increase from a European total of 3 million receivers in 1993 to a total of more than 12 million in the year 2000 of which 5%, with an international subscription, will require European 'roaming' (the possibility of moving, with the receiver, from one country to another and continuing to receive messages without the need for a subscription for each country).

4.6 EUROPEAN PARTICIPANTS

4.6.1 The European Commission

The policy of the Commission has succeeded in establishing the ERMES project whose commercial opening should have occurred in 1993 in some countries. This project is finding increasing approval among those involved in radio paging although its position is not clearly defined with respect to existing services. The establishment of interim solutions, such as the Europage project, and residual capacity on existing networks are such as to delay the volume sale of ERMES terminals by several years.

4.6.2 Standardisation organisations

Many standards are currently in operation but only the ERMES standard is under consideration by ETSI.

4.6.3 The operators

Although the statutory situations of the various national operators are different, they are faced by the same financial dilemma, as follows: how to invest in, and profit from, existing networks, increase the pool of subscribers while preparing for the installation of the ERMES European digital radio messaging system; or, more simply, how to manage the transition from analogue to digital?

4.6.4 The manufacturers

A rapid overview of terminal manufacturers shows the effect of rapid opening up to competition without rules of execution or a European industrial policy. Two giants, the American Motorola and the Japanese NEC, share the European market for the POCSAG system almost totally while capitalising on world markets. Only the Finnish Nokia tends to contain this situation by developing another standard, RDS.

The barriers to entry created by the effect of volume prevent new industries becoming firmly established in this sector. It can be seen that, as far as consumer and professional electronics are concerned, the establishment of a European standard which is not accompanied by strict rules of reciprocity such as access to markets and components will lead to the disappearance of the European industry in this sector within five years.

4.6.5 The distributors

Distribution is essentially professional, but it has developed over several years with the generally low terminal price no longer covering only company subscriptions. Single sales are handled and assisted by consumer distributors.

4.6.6 The users

The major trends in the uses of radio paging can be identified by an examination of the markets of South-East Asia and the United States. These markets, like those of Japan with its 7 million subscribers, Hong Kong and Taiwan, are in advance of European markets by more than ten years. The user's behaviour is increasingly that of the general public while the majority of mobile professional communication requirements are provided by cellular radiotelephone. The second point concerns increasing demand for subscriptions permitting alphanumeric messages to be sent.

4.7 THE UNITED STATES

The United States market is the largest in the world with almost 15 million subscribers. The distribution of subscription types at 65% numeric, 20% 'beep', 20% voice and 5% alphanumeric is indicative of the high penetration of voice frequency telephones which permit numeric messages to be sent easily. The coming years will see a large increase in alphanumeric receivers. A recent development in this sector is the competition brought about by the explosive growth of the cellular market. The low price of services has involved a reduction of operators' margins of the order of 30% and the rate of increase has fallen from 35% to 10%.

4.8 THE FUTURE OF RADIO PAGING

The radio paging receiver in the coming years will become a constituent element of the pocket telephone. The radio telephone or cordless receiver requires a filtering function to avoid the mobile service becoming a burden on its user. To be continuously accessible does not mean to be continuously disturbed but to be informed in real time and at any location. The radio paging function will thus be increasingly integrated into new radio-communication products. Radiotelephones and cordless receivers with this function have already appeared but the major obstacle to this development remains the service surcharge. Operators must offer attractive tariffs so that the subscription cost of two complementary services is not invoiced as the simple sum of the cost of the two separate services.

Two trends thus tend to arise in the development of radio paging; one is towards the consumer market with substantial lowering of prices and aggressive marketing policies and the other is in the sophistication of services and products adapted to certain professional categories.

The consumer radio paging market is one of the major stakes in radio messaging for the coming years. It presupposes greater agreement on proposed products and services, a general lowering of tariffs, terminals which are simple to use and very compact, like a credit card or pen, and a change in the policy of operators in connection with subscription procedures and services in order to adapt to a non-professional clientele.

For their part, the professionals await integrated solutions better adapted to their specific requirements; this includes features such as stock market information, call transfer with voice messages, connection of radio paging receivers to portable or notebook computers or even telepoint terminals, and two-way radio messaging.

It is under these conditions that radio paging will have to find new inspiration, and become both a professional communication tool and a personal fun object.

5 PRIVATE RADIOCOMMUNICATION

5.1 INTRODUCTION

Private radio communication relates to systems and services whose use is primarily limited to a restricted group of users. Even if connection to public networks is authorised, the principal function of these networks remains group communication, of the 'dispatching' type for example. In this large domain, the majority of uses are professional and involve networks such as those for taxis, the fire service, building works, maintenance fleets and also some general public and semiprofessional uses such as Citizen's Band.

Private radiocommunication is the oldest radiocommunication sector and dates from the 1940s. With more than 3.5 million users in Europe in 1992, the installed base shows a slow increase of 5 to 10% per year and it will eventually be overtaken by the public cellular radiotelephone. This apparent maturity should not be allowed to conceal the profound changes which are now affecting this sector; statutory, technological and competitive developments are associated particularly with evolution of the relative positions of the various mobile services.

The typology of this sector will therefore be described by analysing its evolution and examining the specific case of a major user—Electricité de France-Gaz de France.

5.2 TYPES OF RADIOCOMMUNICATION

5.2.1 Citizen's Band

An often forgotten area of private radiocommunication, Citizen's Band, more commonly called CB, has experienced a great upsurge since 1978 in response to a need for conviviality and assistance for people while travelling. In 1984 the European Community Transport Commission reported an estimated 20 million users in Europe. In 1991 the 'European Citizen Band Federation' estimated the number of European users at 25 million of which 2 million were in France; that is four times more than the official number of licences issued by the Ministry of Posts and Telecommunications. This installed base should continue to increase in Europe and this does not include the large population of 'CB-ists' in Eastern European countries where CB is more developed than the cordless telephone.

At the level of service provided, use remains restricted to communication between suitably equipped people within a radius of around ten kilometres. Nevertheless, several channels permit communication without constraints due to proximity.

At the economic level, CB is very attractive since the mean price of equipment is around £100 and the user's licence (subscription and communications) is around £19 for five years.

At the technical level, only 40 channels in the 27 MHz frequency band have been allocated for this service using alternate operation, that is allowing only one of the two correspondents to speak at a time; a manual system permits speech switching.

At the statutory level, CB was the subject of a conflict between user associations and the European PTTs; the latter were seeking to limit the frequencies of a private system which does not bring them any revenue.

5.2.2 Conventional systems

5.2.2.1 *Description of conventional private radiocommunication systems*

In most cases these consist of local or regional single site networks which use a single transmitting and receiving station of limited range. Operation of these networks is regulated by a licence issued by the regulatory authority of the country; in the United Kingdom this is the Department of Trade and Industry and in France the Directorate of General Regulations. This regulation is made necessary by the scarcity of frequencies allocated to this type of service. In France, 32 MHz is devoted to private radio communication networks, with a space between adjacent channels of 12.5 kHz, hence only about 2500 frequencies are available for 450 000 users. Severe congestion problems thus arise and lead to limited availability of licences, non-renewal of allocated frequencies and long waiting lists for communication on the networks. This situation is particularly severe in many European countries such as Belgium, Denmark, Italy and the Netherlands.

The licence conditions of use specify the number of authorised transmit/receive stations per network, the maximum range of the stations (often 30 km), the possibility or otherwise of including stations in the network and limitation of access to the public telephone network.

Operation is in simplex or half duplex mode which permits transmission in a single direction. All terminals thus communicate on the same frequency in one of the following modes: terminal to controlling network base station; terminal to terminal; or network base station to one or more terminals. Successful operation requires a degree of discipline in alternate communication on the part of the users.

In spite of this operational restriction, conventional private networks enjoy great success due essentially to their functionality which responds to the major requirements of organisations and to their low cost. Among the most used functions are call broadcasting, calls to user groups, queue control, priority calls and transmission of data, messages and conditions for remote measurement.

Costs comprise four principal elements:

- 1. The unit cost of the operating licence which is of the order of £200 for 20 mobiles in the United Kingdom

- 2. The purchase price of the terminal which varies from £200 for a bottom-of-the-range terminal installed in a vehicle to £700 for a top-of-the-range portable.

- 3. Purchase and installation of the system, that is the transmitting and receiving station with its operating system which averages £15 000 for a single site system.

- 4. The operating costs which are often the cost of one person full-time.

5.2.2.2 *Division of users*

Users are divided by economic sector and network size. At the European level, a majority of users occurs in services, transport (buses, taxis) and to a lesser extent administration. Fig. 5.1 gives the distribution of private terminals by activity in 1989 in Europe.

A more detailed analysis of the French market shows an appreciably different picture since the health sector is by far the largest with 24% of networks and only 10% of terminals; this is followed by the transport sector with 14% of networks and 21% of mobiles and finally the BTP with 17% of networks and terminals, to cite only the largest sectors. The case of France thus illustrates the large disparities which can exist between European countries, particularly in association with the degree of

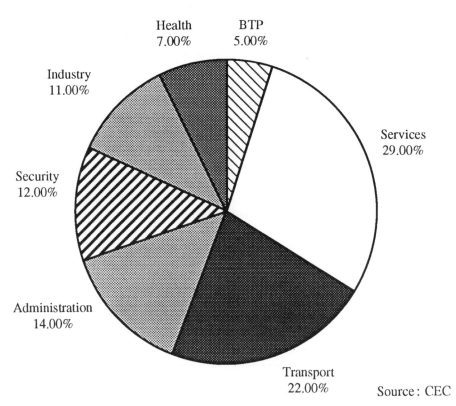

Source: CEC

Figure 5.1 Distribution by sector of private terminals in Europe in 1989

development of the sectors analysed in each country. Also, from an economic and technical viewpoint, it is interesting to analyse the division of private networks in accordance with the number of terminals connected per network.

A net predominance of private networks whose number of users does not exceed five subscribers is evident and this does not justify

Number of terminals per network	Percentage of private networks	Percentage of mobiles
1 to 5	70%	30%
6 to 50	29%	48%
more than 50	1%	22%

Figure 5.2 Division of conventional private networks in accordance with the number of mobile terminals

consideration of a specific infrastructure for the large majority of networks. For these organisations, it would be much more economical to share the investment in infrastructure and recurrent operating costs and to organise these networks into closed user groups or fleets.

Furthermore, from the point of view of spectrum utilisation, allocation of channels spaced at 12.5 kHz for some users enrolled on the service, even over a geographical region of 30 km radius, represents very low spectral efficiency.

5.2.2.3 European characteristics

The principal European characteristics of these networks concern the market, the products, distribution, national factors and the location of manufacturers.

- 1. The European market and the various national markets are characterised by a small increase associated with the maturity of the proposal and the requirements of user organisations on the one hand and frequency congestion on the other. The greater part of the market therefore consists of replacement of obsolescent five to ten year old equipment and the substitution of new products.

- 2. In terms of terminal equipment, portables (weighing less than 800 g) represent at least 30% and at times up to 50% of new sales. With prices growing closer to those of terminals installed in vehicles, portables of increasing autonomy are, like others in the radiocommunication domain, a major product.

- 3. Distribution often remains specialised. It is achieved either directly from the manufacturer to the user organisation or indirectly via distributors who add a large added value as an allowance for the installation of networks which includes engineering, planning, help with installation and rapid after-sales service for the infrastructure and terminals.

- 4. There are numerous national factors. The rate of penetration with respect to the active population is very significant since the Nordic countries (Sweden, Denmark, Norway and Finland), with utilisation by 4–8% of active people, have rates much higher than the majority of other European countries; the national average in Europe is between 2 and 3%. Furthermore, the problems of frequency congestion lead to different behaviour. In the United Kingdom and Sweden, conventional private networks are shared between user organisations which minimise their investments and operating costs in this way but lose in terms of the level of confidentiality and control of the system; these are 'communal networks'. In contrast, in Italy, a third of the spectrum is used illegally by users without a licence.

● 5. Manufacturing locations are shared among manufacturers of the EEC (AEG, Alcatel, Autophon, Bosch, Philips), EFTA (Ericsson and Nokia), America (Motorola) and Asian suppliers. The latter generally penetrate with portable bottom-of-the-range terminals with few functions and low cost. In this market, which has a small increase dominated by terminal renewal, manufacturers having a large part of the market are in a strong competitive position.

● 6. In terms of development, the general tendency is as much technical as statutory; it is the slow migration to new private networks or those operated by a third party; these are shared resource networks or 'trunks'.

5.2.3   Resource sharing networks or 'trunks'

5.2.3.1   *Principle of operation*

Resource sharing networks, also called 'trunks', are private or public networks utilised by an operator to provide a specialised service which permits frequency use to be optimised by sharing among different users.

In this type of network, central equipment allocates a free frequency in real time to the user who requests it, and to him alone, for the duration of the communication. Communication as in conventional networks is half duplex. Furthermore, the central equipment controls a queue of mobiles before allocation of a frequency. The efficiency of trunks can be compared with conventional networks by considering the following simplified example.

Consider three companies C1, C2 and C3 who have the use of three conventional private networks. The networks are single channel and A and B, C and D, E and F belong to companies C1, C2 and C3 respectively. While mobile A of C1 communicates, B cannot communicate. Assume that C and D of company C2 are not communicating. Finally, when E of C3 communicates, F cannot communicate. Let us assume that these three companies subscribe to a shared resource network which has three available channels. Now, if A and E are communicating, one of the four mobiles B, C, D and F can communicate although previously B and F were blocked.

Common use of the resources is thus made and these are better utilised by dynamic channel assignment, a reduction of queues and more efficient management of the spectrum. The figure below illustrates the example described above.

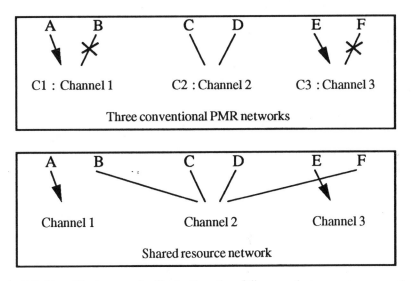

Figure 5.3 Simplified example illustrating the difference between a conventional network and a shared resource network

5.2.3.2 Public trunks

Analogue standards The origins of the three principal analogue standards are: French with Radiocom 2000; British with MPT 1327; and Swedish with Mobitex. In 1992, the total number of subscribers to these public networks remained very low with almost 180 000 distributed as: 100 000 subscribers to MPT 1327 in different European countries; 55 000 subscribers to the trunk service, called the enterprise network, of Radiocom 2000 in France; and 25 000 subscribers to Mobitex mainly in Scandinavian countries.

The MPT 1327 standard was developed in the United Kingdom on the initiative of the Department of Trade and Industry with the objective of reducing congestion on conventional private networks by setting up a national trunk network standard. Since the opening of the British network in 1987, many European countries such as Finland, France, Italy and the Netherlands have adopted this standard. Radiocom 2000, developed in collaboration by the operator France Telecom and the manufacturer Matra Communication, is original in being a standard trunk available on a public network and offering the dual service of a private company network and a public cellular radiotelephone. From the same terminal, which is available for a dual public and company network subscription, it is possible to call a correspondent on the telephone network or in a closed group of subscribers.

Mobitex was developed by the Swedish public operator Televerket and the Ericsson company in response to the growing requirement for data transmission. Voice functions were soon added, however. Mobitex has been exported to North America but at present has been little adopted in Europe, except in Scandinavian countries.

The services provided   Services are very similar on the three standards MPT 1327, Radiocom 2000 and Mobitex. They are accessible from terminals installed in vehicles, portables and transportables; these are generally half duplex and sometimes full duplex in the case of Radiocom 2000 for dual usage of the private network and public cellular network. Access to the public network is not a basic function and remains optional. There are three types of service: voice; data transmission and security and confidentiality functions.

● 1. Voice services. In addition to individual calls, group calling is much used when the destination is a fleet of subscribers, that is a closed group of users defined by the network management which, provided with a terminal called a 'base line', also has a dispatching role. A subscriber can belong to several fleets, up to 15 in the case of Mobitex.

   A priority call to an individual or a group is used in case of urgency. It enables network resources to be freed in accordance with the priority level.

   The networks offer a range of additional services such as call transfer, call recall and conferencing which are not, however, always provided by the operators. These services are rather limited on Radiocom 2000. Mobitex offers a voice messaging service with automatic retransmission to the destination.

   The channel described as 'open' permits free use of a radio channel in case of urgency and a general monitoring of mobiles. These specific functions of private networks such as group calling and priority calling are much used by organisations.

● 2. There are three data transmission services:

   2.1   Predefined messages of 30 to 256 characters, according to the network, permit coded information of telemetry states or changes of state, for example, to be transmitted.

   2.2   Short messages of up to 512 ASCII characters for data transmission on Mobitex provide a mobile messaging service within an organisation or within a fleet.

   2.3   To be of good quality, transparent data transmission via a modem requires an additional error correction function. Mobitex proposes a transmission protocol for messages of more than 512 characters which leaves the format of the exchanged information entirely free.

● 3. Security and confidentiality functions. Authentication of terminals is performed on each call by verification of the serial number. In addition, encryption of the exchanged information is possible but is not offered by the networks. It must be provided by specific terminal equipment.

5.2.3.3 Private trunks

These are to the standards of operated trunks. The total number of terminals in Europe is only 200 000 of which the great majority are in Sweden, a unique European country where the number of users of private trunks exceeds the number of conventional private mobile radios (PMR). The strong general penetration of mobile services in conjunction with a low population density, a habit of mobility and the level of sophistication of users explain this rapid migration to trunks in Sweden and this large advance in comparison with other European countries. In other European countries, the total number of terminals remains marginal in comparison with conventional PMR and progresses relatively slowly. However, congestion of PMR frequencies, encouragement of migration to trunks by statutory authorities, the additional requirement for confidentiality and more elaborate functions all support an increase in the number of users firstly of public trunks and then of private trunks for large organisations after a period of apprenticeship on public networks.

5.2.4 Comparison of PMR, public trunks and private trunks

One of the main problems of an organisation which wishes to equip itself with a private radio communication system lies in the choice between private mobile radio (PMR), a private trunk and a subscription to a public trunk. There are a number of aspects to the choice: the functions offered together with the quality of the service and the total costs including terminal costs; licence costs to operate on the required frequencies; service subscription costs along with the cost of investment in infrastructure and the operating cost. Fig. 5.4 presents a table which compares these three approaches. The cost differences can be large at European level; the comparison is based on the case of one country, the United Kingdom.

PMR thus corresponds to a low cost approach for requirements of limited service, functions and quality. Private and public trunks are similar in level of service. The return on investment of the private network is very rapid in comparison with the public network; the investment per subscriber is comparable to the price of an annual regional subscription to

Criterion	PMR	Public 'trunk'	Private 'trunk'
Services			
Voice	yes	yes	yes
Data transmission	Limited to short messages	yes	yes
Functions	few	many	many
Service quality	average congestion	good	depends on engineering
Coverage	local	regional national	regional national
Price			
Terminals (average price)	£400	£500	£600
Licence frequencies	£10/mobile	N.A.	£5000 (national frequency)
Service subscription	N.A.	regional £20/month national £40/month	N.A.
Infrastructure costs per subscriber	£100	N.A.	£20
Operating cost per subscriber	high	N.A.	low

Figure 5.4 Table comparing the three approaches, PMR, public trunk and private trunk (English example)

the public service. This is the economic justification for future migration of large user organisations from public trunks to private trunks.

5.3 DEVELOPMENT OF PRIVATE RADIOCOMMUNICATION

There are four principal factors in the development of private radiocommunication. Firstly, statutory developments favour the emergence of new specialised operators. New requirements for data transmission lead to the development of new systems such as trunks. European standardisation encourages the establishment of digital standards. Finally, the large growth of the cellular radiotelephone is already an important competitive factor in some European countries.

5.3.1 Statutory developments

Private radiocommunication is an area *par excellence* for the introduction of competition between new specialised operators with characteristics specific to each European country. Today, all countries except Belgium, the Netherlands and some Scandinavian countries have allocated licences to private operators or have initiated the process.

Since 1987, two British companies, Band 3 Ltd and GEC National One, have held a Department of Trade and Industry licence to provide a public trunk service with national coverage. Based on the MPT 1327 standard, these networks offer the whole range of services described in Chapter 5.3.2.2. They compete in the London region with five regional operators.

In France, the Directorate of General Regulations (DRG) allocated six licences in 1990 for the operation of networks in Marseille, Nice, Quimper and Nantes. In 1991 the process was extended to eleven new regional zones. Either the Radiocom 2000 or the MPT 1327 standard must be used. The national territory will be progressively covered by a competitive offer of the enterprise network Radiocom 2000 of France Telecom which is subdivided into 16 regions.

Having authorised the German public operator Telekom to set up its own trunk network, Chekker, the German PTT ministry has allocated six additional licences, in particular to Berlin, Frankfurt and Leipzig. In contrast to the United Kingdom and France, the choice of standard is left to the operator on the basis that economies of scale on single-standard terminals would be small and interconnection of networks is not a priority.

5.3.2 Data transmission

The main question associated with the new requirements bears on the demand for data transmission services. It is important to observe the over-estimate of the anticipated increase in this area; the low utilisation of Mobitex in Sweden for data-only applications is one illustration. This problem is not specific to mobile radiocommunication but common to the whole telecommunication sector. In fact, the cumulative instantaneous data rate of the world's largest public packet data transmission network, Transpac in France, does not exceed 10 Mbps which is equivalent to 150 simultaneous digital telephone conversations!

A substantial increase in mobile transmission services presumes optimisation of the facilities at several levels:

- 1. Portable terminals of the personal computer type must combine the radio function with its transmission components (modem) and error correction with simple ergonomics, reduced bulk and good quality.

- 2. Adequate software must be developed both on the terminals and on central sites to permit and facilitate use in a mobile context.

- 3. The tariff, possibly specific to mobile data transmission, must be attractive in comparison with other methods.

Such conditions can only be achieved in the medium term (five years) with standardisation of small format portable data terminals. This aspect is further developed in Chapter 6.

5.3.3 New systems

5.3.3.1 *Digital Short Range Radio*

Digital Short Range Radio or DSRR is a European response to the Japanese SRR standard for short distance communication. It is the digital version of a possible successor to Citizen's Band and a potential competitor of PMR. Using digital techniques, DSRR permits half duplex voice and data communication over distances of 10 to 20 km.

The technique used is of the trunk type but differs from the trunks described previously. It uses decentralised frequency control and does not necessarily require a central system. Each terminal scans the 79 allocated channels in the 888–890/933–935 MHz band specified by ETSI and uses the first available channel. This very simple procedure avoids continuous monitoring of frequencies while benefiting from the trunk effect of private networks with shared resources and thus offers a better communication quality than PMR. The licensing procedure is consequently as simple as that of CB since the frequency band is common to all European users in contrast to the relatively complex administrative procedure of planning PMR frequencies.

After a period of uncertainty preceding the allocation of frequencies, DSRR should achieve substantial success (several millions of users in Europe) due to the congestion of PMR and CB in some areas and the superior quality of communication, provided that, at the start in 1992, the price of terminals and licences is competitive with these two systems. In this sense, use of the same digital speech encoder as the GSM European cellular radiotelephone will permit DSRR terminals to benefit from economies of scale derived from GSM for this major component of the product.

5.3.3.2 *Digital trunks*

These are being standardised under the name TETRA, Trans European Trunked Radio, by ETSI group RES 6 with several objectives.

- 1. For the Commission of the European Community, it is to promote a European sector which is less developed than in the United States and to improve spectral efficiency by eliminating the allocation of one frequency to one user and by giving preference to trunks.

- 2. For user groups, the benefit of standardisation is associated with expected economies of scale in the price of equipment and greater independence in respect of chosen manufacturer when deciding on the purchase of systems and terminals.

- 3. For manufacturers, standardisation offers both an opportunity to consolidate a competitive advantage at European level by the adoption of a specific technology as a basis for a future standard and the danger of questioning specific developments in which the absence of standardisation would accelerate obsolescence.

In the case of digital trunks, in addition to the usual technical difficulties of standardisation and the different interests involved, an additional question arises: how far should standardisation go? Applications and requirements vary greatly between user organisations, for example a taxi company and the police, and this makes definition of a single comprehensive standard for all applications difficult and undoubtedly inefficient. Under these conditions, several approaches can be considered:

- 1. Standardisation of several standards which apply by sector of use, such as security, taxis etc.

- 2. Standardisation limited to ensuring co-existence of different systems in the same frequency band.

- 3. Standardisation of several basic services and interfaces with services specific to groups of users.

The standardisation timetable provides for finalisation by 1994 and installation of networks to the standard(s) some years later.

5.3.4 Towards cannibalisation by cellular?

This question may at first seem provocative. It is justified by increasing competition between private and cellular radiocommunication networks (see Chapter 6) in countries which are more advanced in cellular techniques such as Norway and Sweden. If the large expected increase of cellular radiotelephones is considered and a ten-fold multiplication of the installed base in Europe occurs in ten years, the risk of cannibalisation is not totally unrealistic. The analogy with private wired networks in organisations is tempting: will a return from private networks managed by the users to public networks with operators who offer virtual private network services occur? This implies the possibility of accessing private network functions on public networks.

Two questions thus arise: have public cellular radiocommunication networks the technical capacity to offer services of the PMR or trunk type? If this is the case, is there a distinct advantage for a user to migrate from a private network to a public cellular network?

The first question involves analysing the capacity of cellular networks for supporting group calls, priority calls, open channels and all the functions of private networks. The infrastructure of the Radiocom 2000 network, which simultaneously offers both types of service, will provide a positive response to this question, even if there are numerous technical constraints.

Under these conditions, the second question becomes essentially economic. Conventional PMR networks are solutions of too low a cost and too low a level of service to be directly competitive with the cellular radiotelephone. Public trunk services have a total terminal and service cost which will not be much less than the equivalent price of a cellular radiotelephone in the medium term. There is, therefore, a medium term threat to public trunk networks. Conventional PMR and DSRR for reasons of cost, and private trunks for users who wish to control their own network, will remain to complement the cellular radiotelephone.

5.4   EDF-GDF CASE STUDY

EDF-GDF is one of the largest private users of mobile telecommunications in France. In particular, the Directorate of Electricity and Gas Services (DEGS), formerly the directorate of distribution for its thousand districts, forms the basic electricity distribution network and has at its disposal a large infrastructure of 600 transmit/receive stations, 15 000 mobiles and 4000 telecontrolled overhead switches. Comparison with respect to DEGS cable communication (approximately 50 000 telephone sets) shows strong penetration of wireless communication, of the order of 30%. Developments in requirements led, in 1989, to the installation of a new network to cover the next ten years. This case study describes the existing situation, the requirements of the DEGS for mobile radio communication and their evolution, the solutions adopted, and finally the organisation for installing the new network with national coverage.

5.4.1   The existing situation and requirements

The current network was designed to respond to requirements in case of a crisis, particularly the logistics of telecommunication to permit rapid re-establishment of the service after severe weather has caused damage to the electrical network. Radio telecontrol, via the overhead switches, permits the defective electrical section to be located. Technical teams on the ground, equipped with on-board mobile sets in the vehicles or portables then re-establish the service manually. They use the same means of voice or data communication to establish the daily operating procedures.

Each of the 85 regional distribution centres has from 4 to 5 transmit/receive stations on its territory to cover 9 districts or branches. Mobiles, on average 20 per district, operate either as general monitors when in service or in half duplex speech mode.

The target plan of EDF, and certain operational constraints on the present radio network, are the origin of the developments in progress. The DEGS has defined its objective as the enhancement of its electrical network by doubling the number of transformers, described as source stations, at the end of the high tension transmission network upstream of the distribution network. The number of sites common to two neighbouring districts is thus increasing as is the need for interdistrict communication. Furthermore, telecontrolled overhead switches (TOS) are increasingly used (2000 new TOS per year) and this creates an increasing requirement for data transmission. Operational problems on the present radio network are associated with insufficiently complete radio coverage when a station is shared between several districts, a limit on the number of mobiles per station and ageing of the hardware.

5.4.2   The new network

After consultation with other directorates of EDF-GDF, particularly the production and transport directorate, which combines the energy, telecommunications, transport and hydraulic production services, DEGS chose a new network optimised for its specific requirements.

On the frequency plan, it uses 24 channels spaced at 12.5 kHz in the 80 MHz band, of which four are dedicated to radio telecontrol of the TOS. Coverage is of the cellular type with a reutilisation pattern of 12 and a frequency reutilisation distance of 120 km.

Each district has access to a station thus ensuring better radio coverage and permitting a greater number of mobiles to be supported, a total of 20 000 on the whole network. The number of TOS will also increase up to 20 000.

The functions offered are provided in three stages which are put into operation according to the phase of deployment of the network. The first permits voice communication and data transmission on the same station, speech communication from the public network to a mobile and communication between two connected districts. The second stage offers in addition the possibility of communicating from a mobile to the public telephone network and connection between any network stations via the public network or a dedicated line. Specific data transmission channels are available at the third stage.

5.4.3 Organisation of the project and the network

To run this development successfully, a multidisciplinary project group has been formed including members of DEGS, the production and transport directorate (to which is attached the telecommunications department), and the design and research directorate. The two principal missions were definition of a frequency plan and centralised planning of the new network. In two years, the following have been achieved: a technical specification, an invitation to tender by potential manufacturers, an analysis of the proposals, the choice of two manufacturers (Motorola and Talco), the technical specification documents and the commissioning of two experimental sites. National deployment will extend over 34 months, up to 1994, concluding five years of intensive development of a network whose planned lifetime is ten years.

Although planning is centralised at project group level, installation project management (propagation studies, station location etc.) is the responsibility of regional transport and telecommunication centres (CRTT), which are centres of radio communication expertise. Hardware and software purchases are the responsibility of the districts, to conform with the decisions of the project group. Maintenance is provided at the first level by the districts and at the second level by the CRTTs.

With an overall hardware, software and operational cost of £48 million over ten years, the new DEGS network is a major realisation of a private radio network in France. It will provide a response to developing requirements for a modest cost of £20 per month per mobile.

6 THE CELLULAR RADIOTELEPHONE

6.1 INTRODUCTION

The cellular radiotelephone is certainly the sector of mobile communications which has most surprised observers and participants by its strong growth since the start of the 1980s. Until the unexpected slow down of the English market in June 1990, two constant factors characterised this sector: the underestimate of the annual increase of subscribers to the service and the infatuation, particularly of financial investors, with projects of high capitalisation and profitability. The years 1991–93 mark a major technological transition from analogue to digital technology, a transition, already observed in the 1980s for public cable communication networks, which will now be applied, not without difficulty or dispute, in the cellular radiotelephone sector.

This chapter, therefore, attempts to answer two questions. Why has the radiotelephone been such a success? Why, under these conditions, has the transition to digital technology remained a problem? To answer these two questions, the following must be examined: the market and its evolution, standards and systems in the world and particularly Europe, the services and products offered, and finally the different participants and their strategies.

6.2 THE MARKET AND ITS EVOLUTION

6.2.1 The world market

Historically, the cellular radiotelephone is Scandinavian. The Swedish public operator Televerket, in the 1970s, appreciated the usefulness and

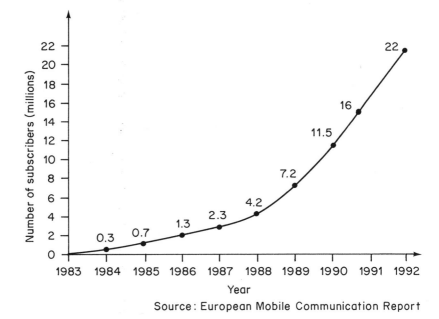

Figure 6.1 Increase of the world total of subscribers to cellular radiotelephone since 1983

potential of radio technology to provide a telephone service in remote geographical regions with a low population density for which the marginal cost of installation of a public cable network would be very high. It needed much persuasion by Televerket to persuade the Ericsson company to initiate the design and development of a cellular system called NMT, Nordic Mobile Telephone, which is common to five Scandinavian countries—Sweden, Norway, Finland, Denmark and Iceland—and whose commercial service started in October 1981.

Simultaneously in the United States, Bell Laboratories, the AT&T research centre, was developing a comparable system called AMPS, Advanced Mobile Phone Service. After several years of testing this was offered as a commercial service in Chicago in October 1983 by 'Baby Bell' which three months later became the Ameritech company following the dismantling of ATT. The cellular radiotelephone is still only ten years old!

Over ten years the number of subscribers at world level was expected to reach 17 million by March 1992. A hundred countries offer a cellular service and most networks have been opened since 1985. The market increase as a number of subscribers continues to remain at a high level, of the order of 30%, and this leads to the conclusion that this service is still at an early stage, that of innovators and leaders of opinion.

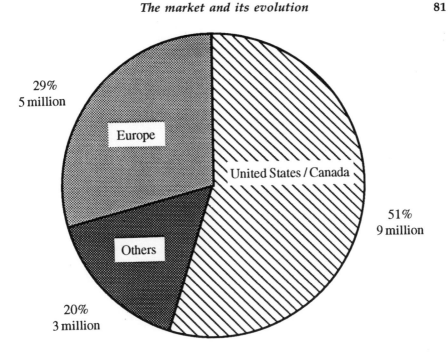

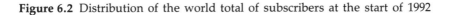

Source : European Mobile Communication Report

Figure 6.2 Distribution of the world total of subscribers at the start of 1992

Certainly, this increase is not uniform and numerous disparities, which will be analysed in Chapter 6.2.3, exist according to country, statutory environment, number of service operators, frequency availability, type of system used, the existence of national manufacturers and the nature of the distribution.

From a geographical point of view, the world market divides essentially between North America, Europe and the Far East.

6.2.2 The European market

The European market without doubt has the greatest differences between countries. In fact, European analogue cellular radiotelephony is characterised by a remarkable absence of uniformity.

- 1. The service started from 1981 to 1987 according to country.

- 2. Many mutually incompatible standards exist: NMT, TACS, C-NET, R2000 and RTMS.

- 3. Penetration rates, defined by the number of subscribers to the service per 1000 inhabitants, vary between 0 and 100 per 1000.

- 4. Statutory environments, determining the number of operators and the type of distribution, vary widely from one country to another.

- 5. The services offered, that is the functions offered on the network, and the supply of products vary widely.

This particularly indicates that, although a user at the end of 1991 can call or be called from a car or portable telephone from 90% of European territory, this same user travelling from Stockholm to Rome by way of Hanover and Paris must not only pay four different subscriptions but must also have four different terminals to be able to call and be called from his car in each of these cities! One terminal corresponds to one subscription and one type of terminal corresponds to one system.

These disparities and the consequent absence of uniformity of services and products for the user at European level have naturally led to a major attempt at standardisation for the digital generation of the radiotelephone. A global approach to understanding the European market is thus difficult since analysis by country, or at least by group of countries, is necessary.

6.2.3   Characteristics by country

To facilitate the analysis, the following groups can be identified.

6.2.3.1   *Scandinavian countries*

These are the pioneers in cellular radiotelephone hardware and their penetration rates are still the highest (see Fig. 6.3). They are characterised by:

- 1. A common standard, Nordic Mobile Telephone, which permits roaming between countries, that is use of the network of one country by a subscriber travelling from another country.

- 2. One operator per country at present, the public telecommunication operator, except in Sweden where a competitive operator, Comvik, remains a minor player.

- 3. Two networks per country:
  a network at 450 MHz essentially intended for car or portable telephones and consisting of large cells of up to 50 km radius;

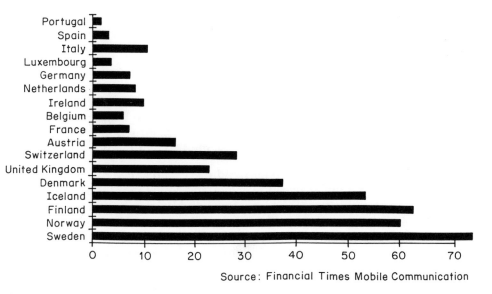

Source: Financial Times Mobile Communication

Figure 6.3 Penetration rate per 1000 inhabitants at mid-1992

a network at 900 MHz, since the end of 1986, which has permitted with a higher density of small cells (less than 3 km radius), the appearance of portable terminals weighing less than 800 g.

● 4. Separate distribution of the service and products:
The distribution of the service (sale of subscriptions, invoicing, after sales service) is provided by the national operator.
The distribution of terminals and their maintenance are provided by well established networks of independent distributors.

6.2.3.2 The United Kingdom

The United Kingdom is the leading country in mobile services in Europe, particularly in respect of the cellular radiotelephone. Although the penetration rate does not reach that of the Scandinavian countries, the number of subscribers is far greater with 1.3 million in 1992. The principal characteristics of this market are:

● 1. A standard, derived from the American AMPS standard, known as TACS (Total Access Coverage System), which was adopted in 1990 by other European countries such as Austria, Spain and Italy.

- 2. Two private operators since the start of the service: Vodafone, a subsidiary of the Racal group, and Cellnet a common subsidiary of British Telecom and Securicor.

- 3. Each operator uses a TACS network in the 900 MHz frequency band.

- 4. There is a range of car, transportable (from 2 to 5 kg) and portable (of less than 500 g) terminals with an increasing predominance of the sale of portable terminals; this represented 80% of the market in 1992.

- 5. Distribution is rather complex and merits a detailed explanation. In granting operating licences to Cellnet and Vodafone, the Department of Trade and Industry (DTI) imposed a service distribution via distinct intermediate operators, called Service Providers. The operator thus concentrates on the technical aspects of planning, installation and operation of the network. The Service Provider is responsible for the sale of subscriptions, invoicing and after-sales service with which the sale of terminals is generally associated. It operates either via a direct sales force aimed at large organisations or via agents or distributors. The Service Provider is remunerated for his commercial performance in service distribution by a percentage varying between 18 and 25% of the overall total of invoices for service, subscriptions and communication. This system thus leads to greater direction of service distribution and greater competition, although in practice the Service Providers offer the same tariffs recommended by the operators. Above all, it permits joint sale of the service, terminal and after-sales service. Furthermore, a bonus mechanism to stimulate demand has been grafted on to this system; it is paid by the operator and the Service Provider and aims to subsidise the purchase of the terminal. In practice, a terminal sold by the manufacturer to the Service Provider for £400 is then sold as part of the associated subscription with a bonus of, for example, £400; one half comes from the operator and the other half from the service provider. If the remuneration from distribution is estimated at £200 for the sale of this terminal by the Service Provider or an agent, the terminal is thus sold to the end user for £200, that is a price less than the purchase price by the Service Provider. Fig. 6.4 summarises the British distribution scheme and the financial flow associated with the example described above.

Frequency availability, competition between operators and between Service Providers, and economies of scale on the terminal market are the key factors of the success of the British market. It should be noted that British industry has not profited at all since the systems provided are produced by the Swedish Ericsson and the American Motorola companies

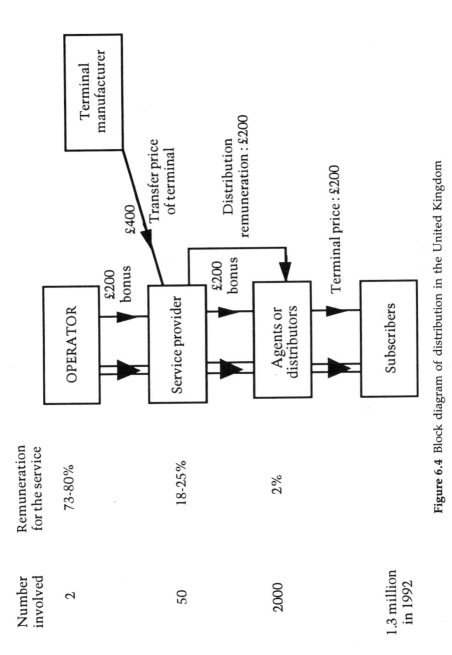

Figure 6.4 Block diagram of distribution in the United Kingdom

and terminal provision is dominated by Motorola and the Japanese company NEC.

6.2.3.3 Other European countries

In comparison with the Scandinavian countries and the United Kingdom, these countries are in the start-up phase of the radiotelephone in terms of penetration rate (see Fig. 6.3). There are diverse standards (NMT, TACS, C-NET, R2000 and RTMS) and sometimes several for the same country: for example, Austria and Spain have NMT and TACS networks and in Italy, the operator, SIP, installed the RTMS network and then, more recently, a TACS network. Only France in this group has two operators: France Telecom operates the Radiocom 2000 network and the French radio-telephone company SFR uses an NMT network. Fig. 6.5 summarises the operator(s), standard(s) used and the number of subscribers at mid-1992 by country.

It should be noted that Germany with heavy deployment in the east, Italy with the TACS network, falling tariffs and terminals which have benefited from the economies of scale of the English market have all led to substantial market increases since 1990.

Furthermore, Eastern European countries are currently in the process of installing analogue cellular networks mainly to the NMT 450 standard. These networks are intended not only to provide a car telephone service but also an alternative or temporary solution to delayed installation of public cable networks. This development is particularly evident in East Germany.

6.2.4 Evolution of the market

Independently of the technological developments which will be described in Chapter 11.3, one of the essential problems of the radiotelephone remains prediction of the market increase. Several methods are possible.

- 1. Evaluation of network capacity, by analysis of the rate of deployment by coverage and the increase of load capacity, in order to quantify the available number of subscriptions; this method is based on the service offered.

- 2. Comparison with models of distribution to professionals and the general public of products having prices comparable to those of radio-telephone terminals; a correction is necessary to take account of service costs (subscription and communications); this method is based on demand analogy.

Country	Operator	Standard	Number of subscribers mid 1992
Germany	Telekom	C-NET	650 000
Austria	PTT	NMT 450	62 000
	PTT	TACS	82 000
Belgium	RTT	NMT 450	55 000
Denmark	PTT	NMT 450	50 000
	PTT	NMT 900	140 000
Spain	Telefónica	NMT 450	68 000
	Telefónica	TACS	73 000
Finland	PTT	NMT 450	155 000
	PTT	NMT 900	160 000
France	France Telecom	R2000	300 000
	SFR	NMT-F	94 000
Iceland	PTT	NMT 450	14 000
Ireland	Telecom	TACS	37 000
Italy	SIP	RTMS	60 000
	SIP	TACS	610 000
Luxembourg	PTT	NMT 450	1 000
Norway	PTT	NMT 450	150 000
	PTT	NMT 900	100 000
Netherlands	PTT	NMT 450	26 000
	PTT	NMT 900	112 000
Portugal	PTT	C-NET	20 000
United Kingdom	Vodafone	TACS	726 000
	Cellnet	TACS	563 000
Sweden	Televerket	NMT 450	245 000
	Televerket	NMT 900	360 000
	Comvik		21 000
Switzerland	PTT	NMT 900	194 000
European Total			5128 000

Source: Financial Times Mobile Communications.

Figure 6.5 The situation by country of analogue networks at mid-1992

- 3. A third method consists of analysing the relation between the increase in the number of subscribers (the number in year N/the number in year $N-1$) and the rate of penetration (number of subscribers per 1000 inhabitants in year N) for each country by year since the start of the cellular radiotelephone; this is the historical method. The curves of Fig. 6.6 show this relation for various European countries and the United States. It can be seen that, although Scandinavia, the United Kingdom and the United States are leaders in

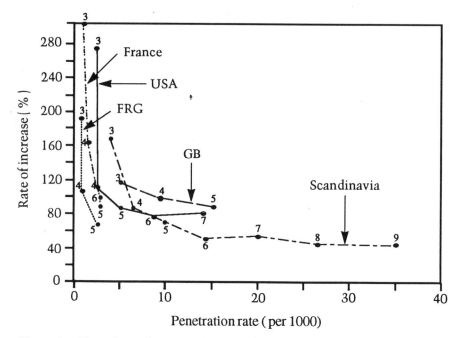

Figure 6.6 The relation between the rate of increase and the penetration rate

terms of penetration rate, five years after the start of the service, the penetration rates in the United States, Germany and France are comparable, about 5 per 1000, as are the rates of increase (between 70 and 90%). Furthermore, since the curves for the various countries represented are fairly close to each other, several general rules of increase can be deduced: up to 0.3% penetration rate, the rate of increase is greater than 100%; between 0.3 and 1% penetration, the rate of increase varies between 70 and 100%; above 1% penetration the increase is still strong and is between 40 and 80%.

6.3 STANDARDS AND SYSTEMS

6.3.1 Analogue standards and systems

The purpose of this section is to compare the principal technical characteristics of European cellular analogue networks. Particular attention will be devoted to the NMT, TACS and R2000 standards while recalling that the TACS standard is derived from the American AMPS standard and a

Japanese 'cousin', J-TACS, used by the Japanese operators IDO and DDI, competitors of NTT.

6.3.1.1 *Principles common to cellular systems*

The various principles common to the cellular systems described from a technical point of view in Chapter 2 are recalled here.
  A cellular system is based on the following:

- 1. Frequency reuse to provide a number of communication channels much greater than the number of frequencies allocated to the system.

- 2. Automatic intercellular transfer or 'hand-over' to ensure continuity of communication when changing cells.

- 3. Automatic location of mobiles within the network particularly to route communications destined for mobiles; these calls are described as 'incoming'.

- 4. Continuous monitoring of communication from the mobile to verify quality and identify the need for a cell change.

- 5. The functions required of all telecommunication networks, operation, maintenance, invoicing etc.

6.3.1.2 *Comparison of the NMT, TACS and R2000 systems*

Rather than making a technical comparison of the analogue standards, attention will be limited to identifying several characteristic elements of the operational systems in Sweden, the United Kingdom and France.
  What are the technical criteria considered?

- 1. The transmitting and receiving frequencies of the mobile, particularly the total width of the available frequency bands, are some of the indicators of the capacity of the system.

- 2. The total number of communication channels.

- 3. The mean number of subscribers per communication channel; this data is specific to each network.

- 4. The spacing of communication channels in kHz; this indicates the spectral efficiency of the system.

Characteristics	NMT 450 MHz	NMT 900 MHz	TACS 2 networks	R2000
Frequencies in MHz				
Transmission mobile	450	890	890	VHF, UHF
Reception mobile	460	935	935	and 900
Bandwidth	4.5	25	2×7.5	3.2 (UHF)
Total number of channels in 1990	10 000	7 000	$2 \times 25\ 000$	6 500
Number of subscribers per channel	24	22	25	30
Channel separation in kHz	25	25	25	12.5
Number of base stations	700	700	2×550	600
Number of switches	12	12	12 per network	STN[*]
Size of cells in km	2 to 50	0.5 to 3	1 to 10	4 to 25

[*] Switched telephone network.

Figure 6.7 Characteristics of the NMT (Sweden), TACS (United Kingdom) and R2000 (France) networks

- 5. The number of transmit/receive stations.

- 6. The number of switches in the network.

- 7. The average size of the cells.

6.3.2 Digital standards and systems

6.3.2.1 *The arrival of digital standardisation*

Europe An increasingly intensive standardisation effort with several objectives has been made in Europe since 1982.

- 1. To remove the incompatibility between analogue systems by the adoption of a single standard simultaneously in European countries from mid-1991.

● 2. To increase the capacity of networks to meet the large market demand.

● 3. To offer services of the ISDN type, NUMERIS in France, particularly with X.25 data transmission services, asynchronous and synchronous transmission from a PC, facsimile or Videotext and services described as supplementary which are comparable to the services which can be accessed with a company telephone such as call diversion, call transfer, call blocking etc.

● 4. Improvement of service quality both in terms of speech quality and confidentiality of communication by the use of digital transmission and signal processing techniques.

The Group Special Mobile (GSM) of the European Conference of Posts and Telecommunications and the standardisation group within ETSI (see Chapter 3.3.3.5) have consequently defined the European standard 'Global Service for Mobile Communications'; this consists of 7000 pages, has been adopted by 17 countries and will form the second generation of cellular systems in Europe.

The United States The American context differs completely from the European one since in the United States the requirement is not to define a new standard to overcome the incompatibilities of existing systems but to increase the capacity of existing analogue systems, particularly in large cities such as Los Angeles and New York where saturation of the existing networks has already been reached.

The Cellular Telecommunication Industry Association (CTIA), the United States cellular radiotelephone standardisation organisation which includes operators and manufacturers, has therefore defined a progressive migration of networks to digital technology in order to provide greater capacity, in practice a multiplication by three. In the coming years, North America will witness a combination of analogue, analogue–digital and purely digital networks and the corresponding terminals. The principle is very simple although its realisation remains complex; in saturated regions it involves replacing an analogue radio channel by a digital one to permit three times as many subscribers to be served.

Japan Japan has hesitated for a long time before instigating digital standardisation; this has favoured the evolution from cellular to micro-cellular rather than a change of technology to increase the capacity of the networks, particularly in large built-up areas. Japanese standardisation is rather similar to American standardisation, apart from some technical differences.

6.3.2.2 Comparison of digital standards

Characteristics	GSM	US (D-AMPS)	JAPAN
Frequency band	900 MHz	800 MHz	800 MHz and 1.5 GHz
Frequency separation	200 kHz	30 kHz	25 kHz
Number of channels per frequency	8	3	3
Coding rate of speech	13 kb/s	8 kb/s	7 kb/s
Compatibility with the analogue system	no	yes	yes
Chanel/MHz	40/80 (half rate)	100	120
Minimum cell radius	350 m	800 m	N.A.
Maximum capacity of speech channels	15	4	N.A.
MHz/km^2	(half rate coding)		

Figure 6.8 Comparison of digital standards

6.4 SERVICES AND PRODUCTS

6.4.1 Analogue services and products

6.4.1.1 Analogue services

Although the services are described here, it is first necessary to recognise that current cellular radiotelephone networks offer essentially a basic service—voice communication permitting calling or being called from a mobile terminal within a national or regional coverage. Additional services are offered by some networks such as call diversion, three-way conferencing, voice messaging and services of the private fleet type (Radiocom 2000). Data transmission is certainly possible, as on the public telephone network, and to be of good quality requires additional interface equipment of the error-correcting modem type. The installation of these services depends not only on the technical facilities of the network but also on the willingness of the operator to develop the functions which are additional to the basic radiotelephone service.

6.4.1.2 Fixing the tariff

This includes three components:

- 1. The unit cost of access to the network, paid in addition to the subscription cost.

- 2. The monthly subscription cost.

- 3. The cost of communications sent and sometimes received which can vary according to the subscription. The average annual invoice in Europe is high and varies between £700 and £800; this is a total average cost of the service and the terminal cost of £1000 assuming amortisation of a £1000 terminal over five years. Fig. 6.9 summarises the tariff components of the Swedish NMT, English TACS and French Radiocom 2000 networks.

6.4.1.3 Analogue terminals

Terminals may be fixed in a vehicle, portable or hand-held.

Terminals fixed in a vehicle generally consist of a radio and logic unit, RLU, mounted in the boot, a remote control at the front consisting of a telephone keyboard, a numeric or alphanumeric display, convenience controls, for example to regulate the volume and give access to specific functions (directory, hand-free operation etc), and a small cableform connecting the remote control to the RLU, the RLU to the vehicle battery and the RLU to the antenna on the roof.

Sometimes additional packages operate as an interface to peripheral terminals of the PC, FAX transceiver or videotext type. As its name indicates, the fixed vehicle terminal supports use in the vehicle, although it may often have a combined vehicle/portable terminal configuration. In other words, a portable terminal is formed by moving the RLU from the

Tariff	NMT 450	NMT 900	TACS (Vodafone)	R2000
Network access	£30	£30	£50	£25
Monthly subscription	£35	£25	£25	£20 to £60(*)
Communication full tariff	£3.60	£3.60	£3.00	£3.60 to £7.30(*)

Note: French and Swedish tariff Conversion with 1SEK = 1F and £1 = 10F.
(*)Regional-national.

Figure 6.9 Analogue cellular network tariffs

boot and combining it with the necessary interconnection and packaging, a remote control and a battery with its charger. These terminals weigh between 2 and 5 kg; autonomous operation is of the order of one hour of communication and eight hours of standby (with the terminal under power and able to receive a call).

Hand-held terminals are enjoying ever increasing success, of the order of 80% of new sales in the United Kingdom and Italy, and are mainly used in countries with 900 MHz networks and small cells of the NMT 900 or TACS type. Their weight is less than 800 g, the lightest are 300 g with their battery and are almost true pocket terminals. Their volume is less than 600 cm^3 and some achieve 200 cm^3. Autonomy varies from 30–90 min of communication and from 6–18 h of standby. The table below summarises the main characteristics of these products.

6.4.2 Digital services and terminals

The description here is more limited since digital services will be progressively commissioned from mid 1992. The following trends can, however, be indicated:

● 1. Services: multiplication of network functions and provision of data transmission services, and a voice service with better speech quality and better confidentiality.

Characteristics	Car telephone	Portable	Pocket
Components produced	RLU remote control cables antenna	RLU remote control battery charger antenna portable pack	Monobloc or flip-flop battery case
Weight	3 to 5 kg	2 to 5 kg	300 to 800 g
Volume	3 to 5 l	2 to 5 l	300 to 1 000 cm^3
Mean autonomy (1)	N.A.	1 h/10 h	45 min/10 h
Mean price without tax NMT Standard	£1 k	£1.3 k	£1.7 k

(1) Autonomy in communication and stand-by.

Figure 6.10 The characteristics of analogue cellular terminals

- 2. Coverage in Europe with the GSM digital standard becomes international, 'roaming' between operators being supported by all signatories to the agreed GSM protocol.

- 3. Tariff: a long-term trend to reduction due to a lower cost per subscriber of the digital infrastructure in comparison with the analogue one. In contrast, in the launch phase, digital services could be more costly than analogue ones.

- 4. Terminals: at the start-up of digital networks, given the increased complexity of the terminals particularly in signal-processing hardware, the terminals will be heavier, bulkier and dearer than the most highly optimised European analogue ones. In the medium term, digital terminals will be comparable to analogue ones due to progress in component integration and assembly technology of these components on to printed circuits.

- 5. Subscriptions and terminals: a memory card called a Subscriber Identity Module (SIM), containing all the data associated with the subscription, permits the user to telephone any mobile GSM, particularly borrowed or hired terminals, in taxis, stations and airports.

6.5 PARTICIPANTS IN CELLULAR RADIOTELEPHONE

6.5.1 Statutory and standardisation organisations

6.5.1.1 *Statutory organisations*

In each country, these have a key role particularly in the following areas:

- 1. Liberation of frequencies, which are often used by the armed forces and microwave broadcast services, for mobile services.

- 2. Choice of new operators.

- 3. Authorisation of terminals.

These three main functions are the determining factors in the expansion of the radiotelephone in a given country. In the United Kingdom, the Department of Trade and Industry (DTI) and the Office of Telecommunications (OFTEL) in allocating a wide spectrum to TACS networks and by favouring the emergence of new operators actively contribute to the development of radiotelephone services. In France, the scarcity and distribution of allocated frequencies is a severe constraint on the Radiocom 2000 service of France Telecom, by limiting the total capacity of the network, complicating capacity increase and involving limited series of terminals.

6.5.1.2 Standardisation organisations

In Europe ETSI plays the major role, particularly in the cellular domain in which one of the major aspects is standardisation of GSM (see Section 3.4.2).

6.5.2 Operators

6.5.2.1 Types of operator

There is a marked contrast between public and private operators.

Public operators offer the whole range of domestic and professional telecommunication services, particularly radiotelephone services; these are the PTTs or ex-PTTs. Their status is evolving rapidly at the present time, and the conditions of their offer of a cellular service are increasingly precisely defined.

Private operators are chosen by a call for tenders by the statutory authorities, often on the initiative of public bodies, the PTT ministry or manufacturers. The conditions of their offered services are precisely defined in a licence of long duration, often several tens of years. The current tendency is towards the creation of international consortia which combine complementary capabilities. The creation of PCN consortia in 1989 in the United Kingdom to provide a mass radiotelephone service on the GSM basis at 1800 MHz illustrates this phenomenon perfectly. Each included a large British group to provide competence in large project management and/or in telecommunications (British Aerospace, STC, Cable and Wireless), one or more telecommunication operators who contribute their experience of operating cable and cellular networks (Pacific Telesis, US West, Cable and Wireless and Telefónica), a commercial supporting company of the 'service provider' type (Millicom) or wide distribution type (Thorn EMI), and finally a manufacturer contributing competence in systems and digital cellular terminals (Matra Communication, STC and Motorola).

6.5.2.2 Evolution

The start-up of GSM in Europe will coincide with a large increase in the number of operators. In seventeen countries which have signed the 'agreed protocol' for setting up GSM, only three have two operators of analogue networks (the United Kingdom, Sweden and France). With GSM, seven countries have already decided to have two or more operators; Sweden has opted for three operators and other countries, particularly in Southern Europe, could have dual operators. In this area again, the United Kingdom is the leader with the allocation of two GSM licences to Cellnet and Vodafone and three PCN (GSM at 1800 MHz) licences. The United Kingdom is thus the European country which has most cellular operators in competition with each other.

6.5.2.3   Operator strategy

There is no general operator strategy, but three approaches can be identified.

- 1. The desire to capitalise on existing analogue networks, without too much investment in GSM; this is the case of Scandinavian operators who have large available capacities with the NMT 450 and particularly 900 MHz networks, and operators who have recently invested or reinvested in an analogue network such as Cellnet in the United Kingdom, SIP in Italy and Telefónica in Spain.

- 2. The desire to deploy a digital network very quickly: this is the case of new private GSM operators such as Mannesmann in Germany, Radiolinja in Finland and Nordictel in Sweden.

- 3. The desire to extend outside national territory; this objective is complementary to the two previous ones and could be common to all operators, but is above all the commercial basis of Racal Vodafone in Europe. This process is broadly favoured by the increasing number of opportunities for obtaining radiotelephone operating licences in the world and the active support of the financial sector attracted by the high profitability of the projects, particularly in cases where the competition remains limited.

6.5.3   The manufacturers

6.5.3.1   System manufacturers

Types of company They are not numerous. Restricting attention to Europe, only five companies have developed a complete analogue cellular system—Ericsson, Matra Communication, Motorola, Nokia and Siemens. However, the number increases if equipment suppliers are taken into account; these are companies which have developed and manufactured part of the system, particularly the transmit/receive radio base stations. This concentration is easily understood due to the numerous capabilities and resources required. The capabilities required of a system producer include the following.

- 1. The capability of managing large projects which often extend over several organisations and geographical areas.

- 2. The technical capability required to contribute to standardisation and expertise in radio, signal processing, real-time software, hardware and mechanical construction.

- 3. The commercial capacity of a partnership to access international markets.

- 4. Financial, technical, technological and human resources.

Evolution Two factors in particular condition the evolution of these industries: the increasing complexity of systems and standardisation. The increasing complexity militates in favour of common use of expertise to share the ever increasing development costs and to minimise development time. It has been brought about by the establishment of GSM consortia and ECR 900 cross agreements firstly between Alcatel, Nokia and AEG; secondly between Ericsson, Matra, Orbitel and Telettra; and finally between Siemens, Philips and Bosch. Standardisation, particularly with the creation of standard interfaces, divides a system into subsystems and creates opportunities for specialist subsystem and equipment manufacturers of one or more elements of the system. Under these conditions, a manufacturer can be the integrator of a complete system of which he provides one element and the other subassemblies are provided by his partners. The Orbitel company is the best illustration—in providing a complete system to the English operator Vodafone of which it is an offshoot, Orbitel provides four components from various sources: the specialised switching from Ericsson, base station control from Matra Communication; the radio transmit/receive base stations and the operating system are its own development.

The strategy of system manufacturers There are world system suppliers, European system suppliers and equipment suppliers.

There are two world system suppliers, Ericsson and Motorola. They benefit from the effects of scale and the benefits of developing to different standards. In spite of their very large resources, at a time of simultaneous migration at world level from analogue to digital with different standards, they had to make a choice of development priority. They capitalise in places where their analogue systems are substantially installed and stimulate digital system development in other regions.

European system producers are rapidly migrating towards digital technology to expand from their existing client base. Their competitive advantage lies largely in the speed of their development and their consequent capacity to modify their system into an exhaustive range of configurations. Alcatel, Nokia, Matra Communication, Philips and Siemens are in this category.

Equipment suppliers provide a particular section of the system. Standardisation of system interfaces and the desire of operators to diversify their suppliers puts them in the position of second rank suppliers. Their competitive strength lies above all in the price. They are Orbitel and Telettra in Europe and NEC and Mitsubishi in Japan.

6.5.3.2 Terminal manufacturers

Types of manufacturer These are much more numerous than system manufacturers. There are about fifty in the world which develop and manufacture analogue cellular terminals although none of them produces a complete range of terminals to all standards. Their origins are very diverse but there are three major types:

- 1. Manufacturers of highly integrated components for terminal fabrication. Motorola and the Japanese NEC, Matsushita and Mitsubishi are in this category. Their competitive advantage lies in the link between equipment supplier and component manufacturer.

- 2. European telecommunication manufacturers, particularly Alcatel, Bosch, Matra/AEG, Nokia and Siemens. Here the strengths are often specific to each company. Among the common strengths are: specific knowledge of European standards by active participation in their definition, expertise in systems and terminals, brand names and distribution networks deployed in Europe, and their direct link with national public operators.

- 3. 'Start-up' companies, of which some have become large, such as Novatel (Canada) and Technophone (United Kingdom) whose major strength lies in the speed of their development and innovations in their products.

Evolution The reduction in the number of standards by the change from ten analogue standards to three digital standards and the increased complexity of development by a factor of 5 to 10 militate in favour of an amalgamation (for example Matra and AEG, Nokia and Technophone) and a reduction in the number of terminal manufacturers. Furthermore, price reduction, miniaturisation and the increasing importance of hand-held units which, in some markets, represent 60–80% of terminal sales, cause a progressive movement of this market from the professional to the general public and will favour entry to this market by consumer electronics organisations which are not yet present.

The strategy of terminal manufacturers Their strategies are clearly linked to their specific strengths. For Europeans, with GSM, it will involve resisting competition by Asian manufacturers by amalgamating or developing co-operation to benefit from the effects of volume and by developing their specific advantages: knowledge of standards, partnerships with suppliers of sets of components, speed of development, skill in distribution and capitalising on European brand names.

6.5.4 Distribution

6.5.4.1 Types of distribution

In general, since the radiotelephone is at present still a professional tool as opposed to a general public consumer item, its distribution remains of a professional kind. There are two types of distribution: distribution of the service, except for the case of the United Kingdom described in Section 6.2.3.2, is realised directly by the operators. This form of distribution covers the subscription, invoicing, often monthly, and after-sales service to the users. Distribution of cellular terminals is very fragmented and often structured very differently from one country to another.

Traditionally the following networks exist:

- 1. Captive distribution networks of terminal manufacturers, particularly manufacturers of telecommunication equipment which combines distribution of radiotelephones, facsimile machines and telephone systems.

- 2 Specialised distribution of cellular radiotelephones. These are often English 'service providers' who, in order to develop a European base and wishing to extend the British approach to the European continent, have set up distribution networks limited to the sale and after-sales service of terminals.

- 3. Specialised radiocommunication distribution. This consists of companies which have traditionally specialised in private radiocommunication (see Section 5.2.2) and have diversified with the appearance of the cellular telephone.

- 4. Automobile distribution of three types: automobile manufacturers who are starting to integrate the car telephone from the start of the design of the car as equipment specific to the car and included when first sold; the concessionaires who treat the car telephone as an accessory sold as an 'add on'; car accessory suppliers to diversify their range, particularly their car radio range.

There are, therefore, very varied participants with very diverse means and ambitions, from the shop which sells five radiotelephones per month to the large national distributor which sells several thousand per month. Often the function of the distributor exceeds mere sale of a terminal. It extends to administrative assistance to the user in subscribing to a network (the distributor plays an intermediate role between the operator and the user), to the installation and programming of the subscription into the equipment, and to the role of adviser in the choice of operator (if there are two), the choice of terminal and its use.

6.5.4.2 Evolution

The distribution sector is evolving substantially for a variety of reasons. Firstly, the marked reduction of terminal prices reduces margins, particularly those of distribution, and often makes forms of distribution using over-qualified personnel obsolete. The strong competition for quality of service to the user requires substantial means and efficient operational structures for rapid installation everywhere in the region (a need for stocks), a quality after-sales service (a need for a means of testing terminals), and a level of advice which reassures the customer (a good knowledge of the service and a relationship with operators and manufacturers). It can be observed that operators, who are concerned to preserve or increase the image of quality of their service, impose more and more rigorous conditions of approval on their installations to guarantee the quality level of terminal installations.

Finally, the structure of distribution of the service is itself evolving greatly. In the United Kingdom, from 1 January 1993, the operators Cellnet and Vodaphone will have the possibility of directly selling subscriptions to complement distribution via 'service providers'. Similarly in Germany and France the GSM operators have announced that they will have non-exclusive recourse to fifteen 'service providers' or service marketing companies.

6.5.4.3 Strategy of the participants

Service distribution In the matter of service distribution, the strategy of the participants hinges on the question: what benefit is there to an operator in using 'service providers'?

This controversial question generates many discussions between supporters and opponents of the British system. Some justify the strong increase in the British market by the existence of service providers. Others reply that the Scandinavian market has developed strongly without service providers and that the subsidised sale of terminals at low prices attracts users who are not capable of paying the communication invoices.

From the operator's point of view, the question of service providers must be posed in terms of the narrowness of the distribution, the quality and added value of the service provided, and the return on investment. Will distribution via service providers be more determined and permit new categories of user to be reached?

Is the service provided, particularly the sale of the subscription to the service and the terminal by the same organisation, more efficient for the final user? Finally, is the increase in system load, that is the increase of users via a 'service provider' type of distribution, sufficiently fast to generate additional revenues and compensate for the payment of 18–25% of the turnover to these service providers? It is the task of the statutory authorities of each country and each operator to respond to this question.

Also, in a context where in the United Kingdom the service providers are experiencing substantial financial difficulties, particularly due to the high level of bonus and major problems of non-payment which they are facing, the economic viability of service providers is to be examined particularly in the context of evolution towards the personal telephone.

Terminal distribution Strategies differ according to the type of distributor and are affected in Europe by the arrival of GSM.

For terminal manufacturers it involves segmenting and controlling the distribution channels by making a good choice of permanent partner.

Specialised distribution of cellular radiotelephones, such as that of the service providers, involves crossing national frontiers to extend over all the countries which have adopted GSM, particularly Germany, France, Italy and Spain.

For specialised distribution in radiocommunication, the aim is to maintain a competitive advantage, particularly with customers in the professional category and regional and local organisations.

For automobile manufacturers, the radiotelephone is often considered to be an important element of the future 'mobile office' which will include, in addition to office automation by communicating PC and facsimile machines, equipment to help with location and navigation.

Finally, one can expect that electronic distribution to the general public will become a major participant in the sector of portable terminals, which do not require complex installation in a vehicle.

6.5.5 The end user

With a mean rate of penetration in Europe of 1% and a maximum of 8% in the Scandinavian countries, the radiotelephone is today still reserved for pioneers. These are small and medium-sized enterprises and the professions which have been the first to subscribe to the service, followed by company managers and commercial executives. It is in the United States, the country of marketing, that most information on user profiles is available. It is observed that 50% of user organisations employ less than ten persons, 50% of users have an annual revenue greater than £35 000 and can be categorised into the following:

- Directors/senior executives: 40%.

- The professions: 33%.

- Middle management: 19%.

- Commercial: 8%.

After a series of trials by several users at high levels in the hierarchy of the organisation, or those who have functions which require much travel, organisations are gradually beginning to equip those categories of their personnel who are the most mobile and require frequent communication with their office. The user is thus gradually changing from the status of an individual professional to that of an organisation. The latter thus controls a quantity of radiotelephones in the same way as telephone receivers or facsimile machines.

In this way the radiotelephone is starting to enter the organisation's telecommunication environment which, as with other means of communication, requires planning, support and management of the service. This evolution in Europe is particularly evident in the United Kingdom where the 'Telecom Manager', who is responsible for the services and means of communication specific to the organisation (such as telephones and data transmission), has an additional tool, the radiotelephone, to manage.

7 THE CORDLESS TELEPHONE

7.1 INTRODUCTION

It is only very recently that the cordless telephone has been considered a major contributor to mobile services. Non-existent five years ago, the cordless telephone now represents 25% of the market by volume in the United States and almost 10% in Europe. New technologies are appearing in which cordless handsets will no longer be restricted to a particular place but will become independent and also serve in public places and organisations. Experiments to test their public use have been made throughout Europe and PABX manufacturers are examining cordless arrangements.

In this chapter the development of the cordless telephone, its applications, the birth of the pocket telephone and the strategies of those involved will be presented.

7.2 THE CORDLESS TELEPHONE (I)

7.2.1 The telephone market

The number of telephones connected to the public network is of the order of 200 million for the European Community as a whole. Fig. 7.1 shows the distribution by country. The total number in Germany, France, the United Kingdom and Italy represents 75% of the European total. The telephone has become a familiar object, but it is only recently that it has been transformed into a consumer item. Telephone distribution is no longer exclusive to the public operator. The subscriber has become free, in most cases, of the obligation to lease his telephone during installation of a new telephone line or a change of subscription by the operator.

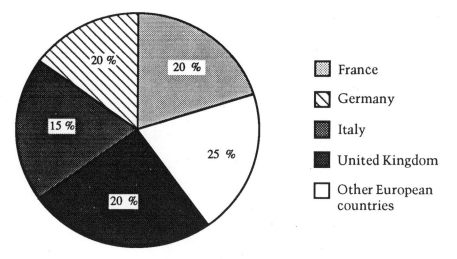

Europe : 200 millions

Figure 7.1 Distribution of the telephone in Europe (all telephones at the end of 1990)

Fig. 7.2 shows the current telephone situation in Europe, the statutory authority in each country and the date of opening the first telephone to competition. The different dates explain why the private market is in its infancy in Spain and Germany but is considerably developed in France and the United Kingdom.

The telephone market has become competitive by diversifying. In most countries it is now possible to obtain a telephone (to rent or purchase) from an agency belonging to the operator or to buy one from an independent distributor.

In all cases of sale, the apparatus is no longer 'the property of the state'; it has become an object which can be chosen. Purchasing criteria result from the proliferation of offers by manufacturers. Consumer choice varies greatly from one country to another. This is associated with the maturity of the market and specific features of national demand. The creation of the single European market does not imply the launching of identical products at European level since national features would be lost.

Many homes today have several telephones located in different parts of the building. A second telephone in a main residence is no longer rare and second homes are more and more often equipped with one. In connection with consumer choice, new purchasing criteria have appeared, such as the design of the telephone, its functions, the price and the brand name. The telephone, in becoming a consumer object, if not a personal one, is no longer merely a tool.

Situation at the start of 1993	Germany	Spain	France	Italy	United Kingdom
Telephone operators	Telekom	Telefónica	France Telecom	SIP	British Telecom Mercury (1)
Telephone liberation	1990	1990	1986	1990	1981
Operator offers (2)	I/S	I/S	I/S	I/S	I/S
G P market (thousands) (3)	600	500	3500	700	6000
Cordless total (thousands) (4)	1200	600	2200	500	2000
1992 cordless market (thousands) (5)	500	100	400	100	600
Cordless standard	CT1/CT1+	CT1	CT0 (F)	CT1	CT0 (UK)
1991 mean cordless price	£250	£250	£120	£250	£80

(1) Establishment of a second operator in 1983.
(2) I : Telephone installation S : Telephone sales.
(3) General public distribution (other than operator sales).
(4) Estimates for approved products.
(5) Approved terminal leases and sales (estimates).

Figure 7.2 The telephone in Europe

With the proliferation of offers and the emergence of consumer choice, the residential telephone market is gradually diversifying into five major areas aimed at different uses; Fig. 7.3 gives the distribution of the market in volume per country.

- 1. The SINGLE UNIT or COMPACT. These basic products, described as 'compacts', are produced as a single unit and consist essentially of a keyboard for dialling, an earphone and a microphone. Normally this range includes only basic functions such as last number repeat and recall.

- 2. The DUAL UNIT. Conventional telecommunication administration telephones are in this market sector. The two parts include a handset, containing an earphone and a microphone, connected by a cord to a base consisting of a dialling keyboard and a bell.

- 3. The MULTIFUNCTION. These are top-of-the-range dual unit telephones (base and handset) having a variety of functions such as ten or more memories (to call the numbers of frequent correspondents more easily), obtaining a line without lifting the handset, a loudspeaker (to amplify signals), a 'hands-free' function (to telephone without the handset in the hand, and being able to converse within several metres of the telephone) and displaying dialled numbers.

- 4. The DECORATIVE. These are gadget telephones with various themes such as telephones resembling cars, bottles or comic strip characters.

- 5. The CORDLESS. The telephones in this last sector are of the dual unit type having special 'cordless' features. The handset is free: it is connected to its base not by a cord, but by radio transmission. In this case the handset also contains the dialling keyboard. The base is connected to the telephone network. It is possible to make and receive calls with the handset in an area which can extend from tens to hundreds of metres around this base. This distance depends on the specification

	Germany	Spain	France	Italy	United Kingdom
Compact	5%	19%	11%	9%	18%
Dual unit	53%	58%	52%	70%	58%
Multifunction	33%	8%	25%	10%	9%
Decorative	1%	1%	2%	3%	5%
Cordless telephone	8%	14%	10%	8%	10%

Figure 7.3 The division of the market

of the cordless telephone and the conditions of use (with open space to the exterior or in a closed region with partitions).

7.2.2 The cordless telephone

Consumer infatuation with cordless devices can be seen throughout Europe, even for limited use in an apartment or studio. In 1993, the cordless total (approved and non-approved) connected to the telephone network is estimated at close to 8 million, with an annual market for 1992 of the order of 3 million. The European total conceals marked variations, as shown in Fig. 7.4.

Although evolving rapidly, Europe is experiencing a significant delay in the use of the cordless telephone, in comparison with the United States and Japan. The total number of cordless telephones in the United States and Japan is 50 and 10 million respectively and the markets are of the order of 10 and 3.5 million.

Cordless telephones are not subject to the same technical type approval specifications, nor to the same policies in respect of these rules in each country. These are two of the factors which explain the large total of non-approved handsets. Market disparities and the mean price of cordless telephones in Europe, as given in Fig. 7.2, show price variations which can reach a ratio of four between two adjacent countries.

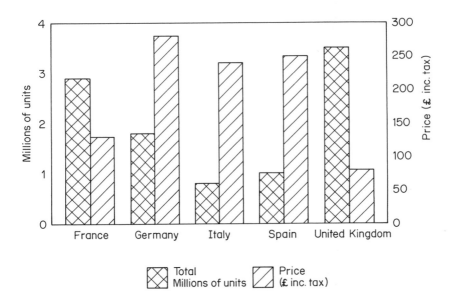

Figure 7.4 The cordless telephone in Europe (total at the end of 1992)

7.2.3 Standards in Europe

Today, the principal standards for cordless telephones (CT) connected to the telephone network are of the analogue type. These include CT0 (United Kingdom) derived from the standard in use in the United States, CT0 (France), CT1, for which the first products were marketed in 1985 from specifications defined by CEPT, and its extension, CT1+, operating over a wider frequency range.

All analogue products to these specifications are described as first generation, in comparison with digital products which are described as second generation and have additional functions. The first digital products did not appear on the market until 1990. The principal standards, whose essential specifications are summarised in Fig. 7.5, are as follows.

Situation at the start of 1991	CTO (UK)	CTO (F)	CT1	CT1+	CT2
Analogue/digital	A	A	A	A	D
Frequency bands	1.6 MHz 47 MHz	26 MHz 41 MHz	914-915 MHz 959-60 MHz	885-887 MHz 930-932 MHz	864-868 MHz
Applications (1)	D	D	D	D/T/px	D/T/px
ETSI status	no	no	standard CEPT	rejected	I-ETS
Hand-over yes/no	N	N	N	N	N
Roaming (3)	N	N	N	Y	M
No of channels	8	15	40	80	40
Transmission (4)	FDMA	FDMA	FDMA	FDMA	FDMA/TDD

Situation at the start of 1991	CT2+	CT3/DCT900	DECT	CDMA
Analogue/digital	D	D	D	D
Frequency bands	864-872 MHz	862-866 MHz (2)	1.88-1.9 GHz	Spread Spectrum
Applications (1)	D/T/px/PX	PX	D/T/px/PX	D/T/px/PX
ETSI status	no	rejected	ETS	no
Hand-over yes/no	Y	Y	Y	Y
Roaming (3)	A	A	A	A
No. of channels	80	64	120	
Transmission (4)	TDMA/TDD	TDMA/TDD	TDMA/TDD	CDMA/TDD

(1) Application areas.
D : Domestic (cordless).
T : Telepoint.
PX : PABX (large capacity).
px : PABX (small capacity).
(2) Maximum : 8 MHz in 0.8-1 GHz band.
(3) Roaming : N : No.

A : Automatic.
M : Manual.
(4) FDMA : Frequency Division Multiple Access.
TDMA : Time Division Multiple Access.
CDMA : Code Division Multiple Access.
TDD : Time Division Duplex.

Figure 7.5 Cordless telephone standards

- CT0 (United Kingdom). This is derived from the standard in use in the United States; the first products appeared in 1984. This is the largest market in Europe with suggested average prices of the order of £80. The cordless total in service today largely saturates the capacity of the system with its eight usable frequency channels. The general image of cordless telephones in Great Britain has been badly degraded by poor radio communication quality due to interference from adjacent radio transmissions.

- CT0 (France). Products corresponding to the French standards lie between the British CT0 and CT1 in terms of complexity rather than price. Only 10 of the 15 channels which the standard permits are in service today. The danger of saturation will not arise for several years although the cordless market shows an annual increase of almost 100%.

- CT1. The first products to the CT1 standard, recommended by the CCITT and defined by CEPT, appeared in 1985. CT1 has been adopted more or less actively by Germany, Austria, Belgium, Denmark, Finland, Italy, Norway, the Netherlands, Sweden and Switzerland. The relative complexity of CT1 causes substantial additional costs for the end user. The price of a CT1 terminal, four times greater than that of a CT0 (UK) terminal, has not allowed the market to take off. Furthermore, the perpetuity of CT1 is now in doubt with the introduction of the new GSM digital radiotelephone system which uses the same band of frequencies.

- CT1+. The CT1+ standard, whose principles are similar to those of CT1, uses a more extended frequency range. Although it has been refused the status of a European standard by the technical assembly of ETSI, Belgium, Germany and Switzerland have nevertheless decided to adopt it. CT1+ supports telepoint application by the addition of supplementary messages. A pilot operation of telepoint CT1+ is in progress, conducted by Telekom, in the German town of Münster.

7.2.4 Cordless telephone participants

7.2.4.1 Statutory organisations

Although the roles of regulation and approval will not, in future, be the responsibility of the public operator in many European countries, the standards in use today have been defined by the telecommunications administrations. These are the statutory bodies which henceforth guarantee that approved products on the market adhere to the technical standards.

Since the frequency spectrum is limited, putting products which do not conform on the market can involve disturbances in their immediate vicinity and pirate communication using neighbouring lines when connected to the network. These phenomena degrade the image of cordless telephones in the eye of the public as was the case in the United States and the United Kingdom in 1985. One of the activities of approval organisations must therefore be not only enforcement of authorised approval but also control of the non-approved telephone market, particularly when the volume of this market is greater than that of approved ones.

7.2.4.2 *Operators*

National operators, by their purchasing policies from manufacturers, have enabled ranges of terminals to be created and costs reduced due to the volume generated.

Cordless telephones are considered specifically by operators due to the limited radio frequency resource. Different national industrial policies and the non-availability of the radio spectrum have led to the formulation of standards using different frequencies; the radio frequency spectrum is still considered to be a strategic resource.

This is one of the reasons why the Bundespost in Germany has attempted to extend its monopoly to the cordless telephone. This attempt had to be abandoned after the intervention of the European Commission.

7.2.4.3 *The manufacturers*

Outside the still dominant position of the manufacturers—national champions according to their operators—the European market is divided by the multiplicity of national standards as shown in Fig. 7.6.

The majority of manufacturers of approved telephones are of European origin. Yet international competition, and particularly that of South-East Asia, is present throughout the grey market of non-approved telephones.

Germany	Spain	France	Italy	United Kingdom
Siemens (G)	Amper (E)	Matra (E)	Italtel (I)	Uniden (FE)
AEG (G)	Philips (E)	Alcatel (F)	Brondi (I)	GPT (UK)
Hagenuk (G)		Philips (F)		
Uniden (FE)				
Philips PKI (G)				

Source of manufacturer: G = Germany, E = Spain, F = France, I = Italy, FE = Far East, UK = United Kingdom.

Figure 7.6 Manufacturers of cordless systems

There has for several years been a concentration of the telecommunications manufacturing sector, with a multiplication of alliances and takeovers in order to prepare for strong competition. In Part 3 the operation of this European policy and the stakes in the battle for the pocket telephone, in which the cordless telephone is one of the major participants, will be examined.

7.2.4.4 The distributors

Figure 7.7 gives the structure of telephone distribution in Europe where the commercial agencies of the operators can be considered as a particular distribution channel.

These are the distributors who act as a catalyst in the development of the telephone. Many years are required to inform and educate the consumer after opening the market to competition. Distributors to the general public must thus take the risk of investing so that the telephone will be recognised not only as a tool but as a consumer object. In this way, the products will gradually be presented in shop windows and seen in operation over a wide area. The cordless telephone, with its up-market image and high price, can be displayed to particular advantage.

These distributors can also cause standards to evolve. By their direct contact with customers and manufacturers, they can directly convey the requests of users to the manufacturers or the public authorities, as was the case in the United Kingdom in 1990. In fact, 60% of the cordless telephones returned by customers for after-sales service did not contain a fault. Concerned at this rate of return, the distributors held a customer survey; this showed that the fault lay with the quality of communication which was rapidly degraded due to the number of telephones and frequency saturation. This was one of the determining factors in the decision to launch new technical generations of products.

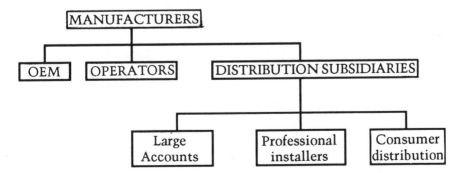

Figure 7.7 Distribution of the telephone

7.2.4.5   The consumers

Telephone subscribers have changed from users to consumers, in accordance with the regulations in force. They now have a free choice from a diversity of offers, but it is only with the application of European directives concerning mutual approval of terminals in Europe and the appearance of standardised products at the European level that the competition will fully take effect.

The size of the cordless section of the telephone market increases with its maturity. Even if the need is not directly felt, as with use in an apartment, the demand for cordless devices increases considerably as soon as the telephone is considered to be no longer an administrative object, but a consumer product. The evolution is such that some experts consider that, at the end of the decade, the cordless telephone will represent half of the market.

7.3   THE CORDLESS TELEPHONE (II)

7.3.1   A triple usage

The cordless telephone will undergo a profound change in the coming years. Its use will no longer be restricted to a given place; the handset, in becoming functionally independent of its base, will assume the status of a mobile terminal.

With the development of new techniques, it will be possible to use a new generation of cordless handset not only near its residential base, but also close to external public telepoint terminals, whose principle is described in Section 7.3.2, and in organisations with radio terminals installed in the private automatic exchange (PABX). It will be possible to use the same cordless terminal for three applications: at home, outside in public places, and in the premises of organisations as shown in Fig. 7.8.

7.3.2   A new concept: the telepoint

The principle of the telepoint or 'communication area' started to germinate at the start of the 1980s from a general report on the situation of pay-phones.

Pay-phones exist today in every country. They permit access to the telephone network from public places. The cost of such a service is limited to the price of communication and use of these terminals does not require a prior subscription to be taken out. Only communication charges are paid

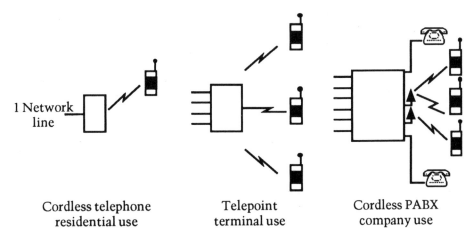

1 Network line

Cordless telephone residential use

Telepoint terminal use

Cordless PABX company use

Figure 7.8 Three uses of cordless telephones

by the user; he only has to provide money. But the problem of vandalism has led some operators to replace coin operated pay-phones by pay-phones incorporating a card reader. Three types of card are mainly used: a bank card, a phonecard and a Telecom card.

- 1. Bank cards serve as a means of payment. The telephone is only one of the possible uses; the user's bank account is then debited. Although very simple in principle, this application does not provide guarantees of security and authenticity. This is the reason why this service is not operated in all countries of the European Community.

- 2. Phonecards can be used only for telephoning; these cards have a limited number of communication charge units in memory (such as 20, 40, 80 or 120). The phonecard is sold at a price corresponding to the value of the charges which it contains by a distribution network chosen by the operator. Each time it is used to telephone, the charge units are decremented in proportion to the communication used.

- 3. The Telecom card is also provided by the operator or an approved distribution network; unlike the phonecard, the communication charges are debited from an account which is associated with a telephone line.

The major failings of pay-phones, most often cited by users, are the lack of hygiene of handsets which are used by everyone, the exposure of public booths, the feeling of obligation to terminate the communication relatively rapidly and the existence of queues for the booths. In fact, although groups of booths are sized according to their location (it is possible to combine several booths in busy places such as stations and airports) and mean rate of occupation queues are inevitable at busy times.

These pay-phone marketing factors, together with developments in radiocommunication services, such as personal paging and the radio-telephone, have gradually established the telepoint concept as presented in Fig. 7.9: 'Following the same principle as the handset of the cordless telephone, it is possible to telephone on condition that one is located, with a telepoint handset, at less than around a hundred metres from a telepoint terminal which is located in a public place and is equivalent to the base of a cordless telephone'.

The telepoint service is thus an access, by way of radio transmission, to the conventional telephone network. This new concept of an access service falls between several categories. It is simultaneously a radio pay-phone or phonecard, a long range cordless telephone and radiotelephone for the general public. It cannot be compared with only one of these existing systems without ignoring some aspect of this new concept.

7.3.3 The cordless telephone in an organisation

Numerous international studies predict that, in ten years, half of company

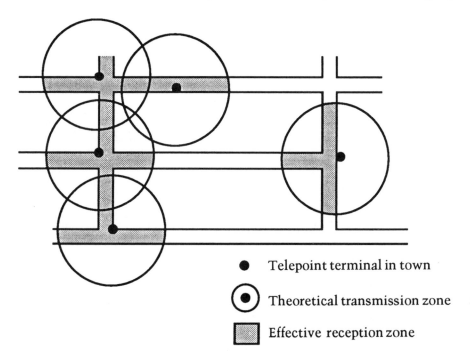

● Telepoint terminal in town

⊙ Theoretical transmission zone

▦ Effective reception zone

Figure 7.9 The telepoint concept in a town

telephones will be cordless. The lack of mobility of communications is most strongly felt within organisations. A telephone number is today associated with a place and not a person. Attending meetings and moving within the organisation, the professional spends less time behind a desk. The number of telephone calls remaining unanswered increases staggeringly. Today, one call in three remains unanswered, and additional PABX services, such as call transfer and storing of calling numbers, do not permit this to be remedied. Analysis of the total cost of these lost calls explains the major interest of organisations in cordless telephone systems where everyone can have a pocket terminal. Unfortunately, the cordless PABX is an expression of unsatisfied demand: commercially available products scarcely exist.

The first cordless systems appeared in 1990 and were based on CT1+ technology but concerned only with small capacities. These products should be compared with sophisticated cordless telephones rather than real automatic exchanges. Developments are in progress using CT2, CT3 and DECT digital technologies on PABXs of higher capacity, but these more complex products will not be commercially available before 1993. The term 'cordless PABX' actually covers three types of system associated with different levels of service, as shown in Fig. 7.10.

The first level of service is established by the installation of terminals within the enterprise. Certain categories of personnel carry a pocket handset which is identified by these terminals. They can thus establish communication from any zone covered by a terminal.

	Outgoing call	Incoming call	Hand-over
case 1	yes	no	no
case 2	yes	yes	no
case 3	yes	yes	yes

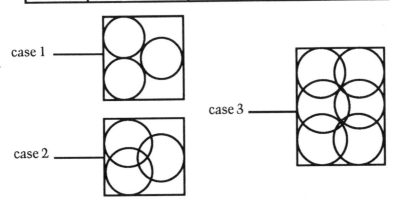

Figure 7.10 The cordless PABX

The second level permits call reception by means of a 'roaming' function. This location procedure, depending on the technology used, is either manual in which a locating message is sent to a terminal using push buttons, or automatic in which the handset transmits this message intermittently. In these two cases, the handset transmits a radio message containing its identification number to give its location. The terminal receiving the message records the number of the handset and transmits it to the central display of the PABX. In the case of an incoming call, the communication is transmitted to the terminal associated with the handset. The terminal attempts to establish communication with the handset. With this second level service, it is not possible to change from one terminal to another in the course of communication in order to improve the quality of communication or move about. The 'hand-over' function does not exist.

The third level of service does have a 'hand-over' function. It is possible to move within the organisation, to call or be called from any terminal point and continue to move while communicating.

Market studies have shown the acceptability of such applications in organisations with a surcharge of the order of 25% per installed PABX line. It is estimated that 20% of all telephones installed in small systems, micro-exchanges, hybrid systems and dedicated systems will be cordless by 1996. Resistance to the introduction of cordless telephones in organisations will be greater for automatic exchanges with a large number of network lines, but synergy with telepoint could activate this sector from 1994. In fact, 90% of potential cordless terminal users in an organisation would take out a telepoint subscription for the same terminal as that at their work if the technical specifications of the two services would permit it.

7.4 TELEPOINT

7.4.1 Technologies in competition

The telepoint is a new generation of mobile services which will develop in the coming years in Europe. The standard most suited to this innovative application is under discussion. Competition involves the various standards summarised in Fig. 7.5.

- CT1+. A CT1+ telepoint site is being tested at Münster (Germany) by Telekom. Also, some manufacturers such as Autophon, Hagenuk and Dancall are offering small systems of 2 or 3 network lines which are adaptable for PABX use. This CT1+ technology is used in its telepoint application only for public testing of the concept of the service since products for this technology are already available. The specification of CT1+ will not permit the objectives of the pocket telephone to be achieved at the competitive price required to develop the telepoint market. CT1+ products will remain conventional cordless devices, possibly for professional applications with intercommunication (communication between two handsets registered on the same base) and transfer functions.

- CT2. Of British origin, CT2 has developed since 1989 to take account of the requirements of different operators and to become a European standard. The first products conforming to this standard appeared in the second half of 1991 in Great Britain, Germany, France and Singapore. Some manufacturers also offer small capacity PABXs. The next two years will be decisive for CT2 in its battle with the DECT standard promoted by the European Commission.

- CT2+. Developments are currently in progress in Canada by Northern Telecom which will add signals permitting 'hand-over' and 'roaming' to the CT2 specification. This development, however, guarantees upward compatibility of the system with conventional CT2 handsets and permits optimum use of the PABX application.

- CT3. CT3, also called DCT900, is a system developed by the Ericsson company for an application of the PABX type. This new standard, like DECT, uses the TDMA principle but in the 900 MHz frequency band. However, ETSI refused at the start of 1991, in spite of intense lobbying by Ericsson to adopt CT3 for the interim European Telephone Standard I-ETS.

- DECT. The DECT specifications were published in 1992 for definitive adoption as the European cordless standard and the first products produced in volume should appear in 1994. The next five years should see competition between cordless CT2 and DECT with a more professional position in PABX applications for DECT rather than CT2 which is oriented towards the general public and residential, telepoint and small PABX applications.

7.4.2 The British telepoint

The telepoint concept originated between 1970 and 1980 in many countries but it was in the United Kingdom, on the initiative of the Department of Trade and Industry (DTI) and several manufacturers, that the concept became reality. The CT2 programme, presented in Section 9.2.1.1, led to the opening of the service in 1988, but events took command. The British telepoint was in stalemate.

Three operators (Callpoint, Zonephone and Phonepoint) operated their service for nearly a year and installed between 5000 and 10 000 public terminal points on major motorways, in the centre of London and several large towns; but there were only a few thousand subscribers at the start of 1991. This setback had several causes, which were primarily technical delays but also major marketing and commercial errors.

In the first place, the setback was attributable to the delay in formulating the specification. Amendments to the 1989 standard were not published until a year later, thereby delaying product development and marketing. The manufacturers, before offering the first public terminal points and handsets conforming to the standard, had nearly a year's delay on the initial DTI project. The first CT2 products conforming to the final version of the CT2 specifications, called CAI, received approval only during the second half of 1991. The operators thus preferred to delay installation of terminal points and marketing of handsets in order to avoid replacing interim products only a few months after installation and marketing. The operator BYPS is the only one not to have opened its service, RABBIT, with interim products. Its strategy was focused from the start on definitive products, having preferred to lose the advantage of being the market leader. The RABBIT service was expected to open at the end of 1991, two years after the other operators, with the appearance of approved CT2/CAI products.

The second error, marketing and technical, was not to have capitalised on the multiplicity of uses of handsets by limiting marketing to telepoint handsets without offering residential terminals.

Telepoint communication has developed from the existing radio-telephone service. Launching a new concept is always difficult, as the public still likes to cling to a known concept or product. The special radio-telephone distribution structure in the United Kingdom with 'service providers' subsidises the sale price to the final customer. The price of radio-telephones thus becomes less than their real cost. The telepoint handset, presented as a 'pedestrian's radiotelephone', thus finds itself in direct competition with the radiotelephone in terms of price. Fig. 7.11 presents a comparison of the tariffs in use for world telepoint services.

Apart from the choice of service names which causes confusion (no-one knows the difference between Callpoint, Callzone, Phonepoint and Call-phone), the major factor which has caused the initial setback to telepoint in the United Kingdom is the inadequacy of the services offered due to uncontrolled competition. Three of the four operators, driven by competition, opened as soon as operating licences were obtained, after installation of only a few tens of scattered terminal points in London. In this situation of savage competition, the subscriber must find a terminal point which corresponds to his subscription (Callpoint, Zonephone or Phonepoint) since 'roaming' between operators is not yet operational. This roaming function permits a subscriber to one operator to use the terminal points of another competing operator, particularly in regions where his nominal

Country	Germany	Netherlands	France	UK	Singapore	Hong-Kong	Thailand
Service name	Birdie	Greenpoint	BiBop	Rabbit	Callzone		Fonepoint
Entry fee (£ inc. tax)	23		20	15	3	5	14
Monthly subscription (£ inc. tax)	3	1.8	4	6	5	11.5	7
Surcharge/basic tax (%)	60%	100%	100%	100%	100%	100%	100%

Figure 7.11 Telepoint tariffs

operator is not represented. The 'interoperator roaming' function, for technical, and above all political, reasons will be implemented only from 1992 on a case-by-case basis.

The savage competition between operators means that the majority of terminal points are installed in the same places in order to obtain the maximum traffic. The three operators' terminal points are thus found at the same place, while a few hundreds of metres away the service is not offered.

The combined effect of these errors led to a rationalisation of the competition between telepoint operators. The financial problems of Ferranti in the United States led them to offer Zonephone for sale. Only one operator remains optimistic of the future of mobile services at this time; he is Hutchison, the director of the BYPS group. The other operators have dropped out.

7.4.3 International projects

7.4.3.1 MOU CT2

The involvement of European operators in the Telepoint project since 1987 was formalised in March 1990 by the signature of a formal agreement CT2/CAI, a genuine frame of agreement defining the European regulations for the installation of pilot telepoint sites during 1991/92 in the various signatory countries. Roaming between operators would also be in place later. A handset, having obtained approval in one of the countries of the European Community, could be sold in the other countries and would operate on all telepoint systems.

This agreement, initially involving eight countries, today includes about a dozen countries among which are the United Kingdom, France, Germany, Spain, Italy, Belgium, Portugal, Finland and the Netherlands. Singapore has since joined the consortium and should open the first CAI service in the world.

Numerous countries are also evaluating the CT2 standard without actually signing the formal agreement. Experiments have also been made in Canada, the United States, Hong Kong and Australia. But the most active countries remain Germany, France and the United Kingdom.

7.4.3.2 Germany

The policy of the Bundespost has developed strongly in recent years. Considered as one of the most commanding operators in Europe, Telekom, following its change of status to a private company, has followed a strongly competitive market policy. This is the reason why the initial offers to tender for the provision of pilot telepoint sites by manufacturers have not been restricted to a given technology; the market can decide for itself. Of the three pilot sites proposed after consultation in January 1990, only two have been retained: one at Münster using CT1+ technology and the second at Munich using CT2/CAI. The third at Dortmund, which was to be a hybrid of the two, has finally been cancelled.

The CT1+ system opened at the start of 1991, but the delay in the development of CT2/CAI terminals has put back the opening of the CT2 network to September 1991. However, the first stages of marketing have already started at Münster and clearly show a large interest by the general public and professionals for pocket handsets at prices in the vicinity of £150. This objective cannot be achieved by choosing CT1+, a rather expensive technology which is little integrated, but the manufacturers of CT2 must now prove that this digital technology can allow the objectives not only of size (less than 200 cm^3) and weight (less than 200 g) but also of price (less than £150) to be achieved in 1993. The commercial objectives of Telekom are to achieve a pool of 500 000 subscribers in 1995 and 2 million in the year 2000.

7.4.3.3 France

Since the start of 1990, France, with Germany, has become the leader of the CT2 telepoint project, causing the CT2/CAI standard to develop by incorporating new functions and eliminating weak points. After a technical evaluation of the British project, the French telepoint system Pointel was based on the CT2 standard and included improvements relating to authentication of the subscriber, protection of the network against fraud and additional services.

The system has been dimensioned from the start on the basis of a centralised national network. The British telepoint networks are based on the principle of a black list of handset codes which are not authorised to telephone from telepoint terminals. The file is downloaded daily or weekly to the public terminal points to prohibit access to the network by unauthorised handsets.

In contrast, the French network uses the principle of the 'white list' to provide greater security against fraud. All public points are connected via a national X.25 network to a central file containing all handset numbers whose access to the network is authorised. Before establishing communication, the system verifies the authenticity of the handset in real time. This procedure provides greater security against attempts at unauthorised connection of handsets to the network.

The wishes of subscribers have also been taken into account with the introduction in the CAI standard of two additional services on the initiative of France: remote registering and incoming call facilities. Remote registering will permit a new subscriber to register on the system as a new subscriber more simply than previously during his first communication from a Pointel point. It will be sufficient for him to enter a specific code of several numbers on the keyboard of his handset, while the initial procedure used in Great Britain requires about forty characters to be entered with no allowable error.

Incoming call facilities in the telepoint application are also a French proposal arising from a subscription to the CT2 standard. This additional service allows calls to be received by means of a manual location procedure which has not been included in the British network. This clause was introduced by the DTI to limit direct competition between telepoint and the cellular radiotelephone. Incoming call facilities will thus be offered on the Pointel system as an additional charged service.

The basic service permits only outgoing calls from public points. To be able to be called, the subscriber must locate himself close to a point and press a given sequence of keys; the system recognises the number of the handset and informs the central file of its location. In the case of an incoming call, the system will route the communication to the point at which the handset is located and also to adjacent points.

7.4.3.4 Singapore

The Singapore operator joined the initial group of telepoint operators six months after signature of the agreement, but it could put the first CAI system into service. One of the important characteristics of this country is its strong penetration of radio paging services with more than 10% penetration rate compared with a maximum of 2% in Europe (Norway). The success of radio paging will provide strong commercial synergy with telepoint from the start.

7.4.4 Participants in telepoint

7.4.4.1 The European Commission

By its executive action in the formulation and promotion of DECT in the face of other possible choices, the European Commission has shown itself to be concerned from the start for the survival of European industry and the right of consumers. This policy of decentralisation and co-ordination has born fruit in large development projects such as JESSI (the European component sector), RACE (long term research and development projects in the European telecommunications sector) and DRIVE (research programmes in the vehicle domain).

But a division appeared between the experts of the Commission on one side and some operators and manufacturers on the other when in 1987 in a technical/political struggle over the choice of coding for the radio transmission, TDMA was preferred to FDMA for the European cordless multifunction telephone DECT. However, after two years of lobbying against CT2, the European Commission has chosen to let the market alone determine the success or failure of CT2 in the face of DECT.

7.4.4.2 Statutory organisations

ETSI defined DECT as the European Telephone Standard (ETS) for the multifunction telephone and has adopted CT2 as the interim standard (I-ETS) while rejecting other proposals for adoption as I-ETS from CT1+ and CT3 (DCT900). In permitting the co-existence of two standards, ETSI demonstrated its major supervisory role and left the development of the telepoint market to operators and manufacturers.

7.4.4.3 Operators

In signing a memorandum on telepoint CT2 in April 1989, the majority of European operators had a pragmatic approach. After a false start in Great Britain, it is the French and German operators France Telecom and Telekom who have taken the European leadership in these projects.

These are the strategies of operators who will to a large extent determine the success of telepoint although the public market must be supported by private residential applications and professionals with PABXs.

7.4.4.4 Manufacturers

The original manufacturers of CT2 were British. The initial group which signed the formal CT2/CAI agreement with British Telecom included STC, Shaye, Orbitel, GPT and Ferranti. The structure of all these companies has evolved over five years. The telecommunications division of STC has been bought by Northern Telecom, Nokia has withdrawn from its holding in Shaye, the Orbitel subsidiary of Racal has been 50% taken over by Ericsson, GPT has become a subsidiary of Siemens, and Ferranti has had numerous financial problems. The whole telecommunication sector has been disrupted and these significant developments and the systematic disappearance of the British industry have cast doubts on the viability of CT2.

The arrival at the start of 1991 in this market of Motorola pocket products, the first worldwide manufacturer of radio terminals, and Shaye together with the announcement of developments from Sony and other manufacturers will re-establish a new order to revitalise CT2.

7.4.4.5 *Distributors and consumers*

A comprehensive assessment of the reception of telepoint by distributors and the public can be made only two years after the commercial opening of the service, that is in 1994. In fact, the market response to the introduction of new concepts takes time to overcome the stage of a gadget for innovators.

7.5 THE EVOLUTION OF CORDLESS TECHNOLOGIES

The principal development of the cordless telephone remains the delocalisation of communication. The handset is no longer uniquely linked to one base, and hence to a given place, but to a person. Having an identifying number, he can access the system from a multitude of terminal points. This new product is not fundamentally innovative but will evolve with the appearance of public services such as telepoint and private ones with cordless PABXs. This new marketing approach will give this handset the status of a personal communication device and bring it closer to the PCS products of the year 2000 described in the third part of this book.

8 MARINE, SATELLITE AND AIRCRAFT COMMUNICATIONS

8.1 MARINE COMMUNICATION

The requirements for marine communication have been known for a long time. They include coastal communication, emergency communications and the need for meteorological information. Potential users are either large pleasure boats (greater than 8 m), of which the total exceeds a million in Europe, or commercial ships which include 25 000 merchant ships, 100 000 fishing boats and about a thousand ferries of various types. Many means of communication are used including VHF, HF, cellular radio-telephone close to the coast and Inmarsat standard A for communication on the high seas.

8.1.1 HF and VHF communication

These services are provided by coastal stations to ensure almost complete coverage of the European coast up to 50 km offshore. To call a boat, the call is established on a free channel by the coastal station which manually establishes communication with the number requested. The system also permits incoming calls, reception of marine meteorological information and provision of an open channel for emergency calls and appointments. Tariffs are of the order of £3.00 for 3 minutes on VHF and an average of £7.00 on HF. Commercial ships are generally equipped with one VHF and HF terminal and make several telephone and telex calls per day.

8.1.2   The cellular radiotelephone

Good propagation conditions at sea permit ranges of up to 100 km to be achieved. The cellular radiotelephone is most used on pleasure boats and commercial ships in Scandinavia using the NMT standard which is common to various countries and the United Kingdom (the TACS standard). The main advantages for users are lower communication costs, compared with other means, and combined use of terminals at sea and on land (see Section 6.4.1.3).

8.1.3   The INMARSAT standard A service

Inmarsat is an international organisation consisting of 60 members, mainly PTT administrations, responsible for providing marine satellite communication services. This system uses 8 satellites, which are not all dedicated to maritime applications, and 20 earth stations (antennas) which are spread over the 5 continents and give access to the public cable networks. The services offered are mainly telephone, telex, data transmission and facsimile. Terminals, of which there are 12 000 in the world, generally give access to a telephone line and a telex line. Although this service is known for its reliability, penetration remains limited by the high cost of terminals and communication (£15 for 3 minutes).

8.2   MOBILE TERRESTRIAL COMMUNICATION BY SATELLITE

8.2.1   Competing services

Other than the marine communication described previously, the principal application area of mobile satellite communication is road transport. Optimisation of the management of lorry loads and deliveries requires a system of vehicle location and messaging between moving lorries and fleet management. This service can be provided by a satellite system with wide coverage over Europe. To complete the service, the system must be capable of integration with a data base showing maps of Europe for easy location together with a data-processing system internal to the transport company for efficient control of the messages transmitted by the lorry drivers. Market studies predict one million users of such services in the year 2000. Without doubt demand will be stimulated by competitive offers since three systems are in competition on the market—Inmarsat, Euteltracs and Locstar.

8.2.1.1   INMARSAT standard C

The Inmarsat standard A service is used for terrestrial applications which support voice communication, telex and high speed data transmission in isolated areas. The standard C service is used for message transmission up to 32 000 characters. Vehicle location is achieved by combining terminals with Global Positioning System (GPS) terminals. Furthermore, portable terminals in a 'suit-case' version permit use out of a vehicle. Operational from the end of 1990, the standard C service provides services for messaging, location for security reasons or management of fleet vehicles and remote data collection. Its main strengths are technical, with easy interconnection to public cable networks, and commercial, having the support of the 60 members of the Inmarsat organisation.

8.2.1.2   EUTELTRACS

This service, whose original technology is American, is provided via Eutelsat, the European organisation of continental satellite services. It permits location of equipped vehicles to an accuracy of 400 m and provides a communication service for messages which are reformatted or in free format of 2000 characters. The terminals consist of three sections: an omnidirectional antenna located on the roof of the lorry, the radio and logic unit weighing 8 kg and mounted within the vehicle and the keyboard to permit composing, sending and reading messages. Operational in 1991, this service is marketed by European operators who are signatories of Eutelsat. The terminals are provided by Alcatel-Qualcomm, a subsidiary company of European manufacturers and the American designer of the system.

8.2.1.3   The LOCSTAR project

Created on the initiative of the French national centre for space research (CNES), Locstar was an association combining 50 European participants of many origins including manufacturers in the space sector, European operators, transport vehicle manufacturers and bankers. This ambitious project proposed to offer a service of high accuracy location (less than 100 m) and communication of short messages of 100 characters with two satellites. The very high risk of the project obliged the participants to lower their ambitions by limiting the space segment to one satellite and then to abandon the project completely in 1991.

8.2.1.4 Overview and uncertainty

By using several simple criteria such as coverage, service functions, date of entry to commercial service, price of the service and terminals, and type of marketing, Fig. 8.1 summarises the principal characteristics of mobile terrestrial satellite services.

Entry into service of the two systems described above and termination of the Locstar project should not mask the substantial uncertainties associated with the satellite radio paging and messaging market. These are as follows:

● 1. Commercial uncertainties: large scale acceptance by the transport sector of a sophisticated data communication system installed in lorries remains uncertain in spite of the attendant productivity benefits. The rate of penetration of such an innovation is particularly difficult to predict. It should be observed that in the United States, very much a country of long distance road transport, American equivalents of the Euteltracs and Locstar systems totalled only approximately 20 000 subscribers after three years.

Criterion	Inmarsat standard C	Eutelsat	Locstar (1991 Project)
Coverage	Europe	Europe	Europe
Commercial opening	End of 1990	Start of 1991	End of 1992
Type of service	Data transmission (32 K char)	Positioning (<400 m) Messaging (2 K char)	Messaging (2 K char)
Terminal price (indicative)	£5 k	£3 k	£1.5 k (objective)
Service price	70 p/1000 characters	30 p/message	30 p/message
Marketing	Operators Inmarsat signatories	Operators Eutelsat signatories and Service provider	Locstar subsidiaries in Europe
Terminal provision	9 manufacturers	Alcatel Qualcomm	3 manufacturers

Figure 8.1 The principal characteristics of terrestrial mobile satellite services

- 2. Statutory uncertainties: in spite of the presence of operators in the three projects, their motivation for rapid commercial opening of the service together with the statutory process of authorisation can vary greatly between countries according to the service. The danger is that, at the commercial opening and for a long subsequent period, a service would not be available in all European countries, as was the case for other satellite services in the past. Such an absence of uniformity at the European level would be highly detrimental to the development of a genuine cross-frontier service.

- 3. Competitive risks: the ultimate service of location and associated messaging could be provided for the user by other less costly means even if their operation were less specific to road transport. The GSM European cellular radiotelephone provides a short message service and permits mobiles to be located. Even if its European coverage was never as extended as that of a satellite, its useful coverage for road transport is very good and includes motorways and major arterial routes. Furthermore, the price of GSM terminals will be very much less than the price of satellite communication terminals, an important factor for a potential clientele which is very sensitive to price.

- 4. Financial risks: these follow from the previous uncertainties, are higher for a greater initial investment and are irreversible. This is without doubt the reason for the failure of the Locstar project.

8.2.2 New projects

Targeting the world market for messaging, radiotelephony, mobile data transmission and radiolocation, technically innovative projects are flourishing at the start of the 1990s.

Launched by Motorola, with much media attention, the Irridium project provides for the creation, in 1996, of a world network consisting of 77 low orbit satellites serving, in particular, regions of low population which are difficult to access by terrestrial networks.

Each satellite would provide coverage of a group of 37 macrocells each of 300 km radius. The low orbit of the satellites permits the power of the mobile terminals to be reduced and rapid launching of a new satellite in case of breakdown of one of them. The capacity of such a network, according to Motorola, would be several millions of users.

This project is very ambitious and its difficulties are as much technical as statutory, commercial and financial; it will require a great deal of international co-operation by operators and manufacturers to be put into operation. It is the first attempt to define a mobile service with world coverage.

About ten other American and European candidates have been declared including Inmarsat Project 21, CNES project S-80 and Loral-Qualcomm Globalstar. The 1992 World Radiocommunication Conference (WARC) allocated the 1610–1626.5 MHz frequency band for this type of application thereby opening up the possibility of allocation of frequencies and operating licences by country.

8.2.3   Case study: analysis of the profitability of a satellite communication service for a transport company

This case study is the result of an experiment performed over three months in 1990 with the Euteltracs system in the Netherlands by the Koning Company for Hartman Mobicom B.V. and four Dutch road transport companies. This realistically sized test was carried out with thirty terminals mounted on lorries travelling in Western Europe under normal operating conditions and permits an initial cost/benefit analysis of the use of such a system to be achieved.

Firstly, there are non-quantifiable advantages in the use of a satellite communication service including time savings for the managers of lorry fleets because of a more efficient method of management, a better level of information transmitted to customers on the location and arrival of their cargoes, and a calling facility in case of problems on the route.

Costs depend on the purchase of mobile terminals and a personal computer provided with specific software and a database of maps for fleet management, a subscription to the Euteltracs service and the cost of satellite communication.

Costs		Savings	
Amortisation of mobile terminals	£126 000	Telephone calls	£30 000
Amortisation of central control equipment	£3 500	Drivers' time saved	£67 500
Mobile communication	£30 000	Operating time saved	£27 000
Central site communication	£1 500	Optimisation of deliveries	£52 500
Total	£161 000		£177 000
		Net saving	£16 000

Notes: Annual base in florins (1 FL = 30 p) 30 travelling mobiles.

Figure 8.2 Costs and savings associated with the use of a satellite service

Financial gains result from saving drivers' time due to a reduced number of stops for fleet management, economies in telephone communication which is replaced by satellite communication, operational gains in the routeing of cargoes (fewer detours of lorries and more rapid execution of customer requests) and optimisation of cargo transport by the fleet (grouping of adjacent deliveries). Fig. 8.2 summarises the details of the charges and the gains realised. If the net financial gains are smaller, they will nevertheless provide an immediate profitability on a strictly financial basis.

8.3 AIRCRAFT COMMUNICATION

8.3.1 Principle of operation

In this section air–ground communication is considered in order to permit commercial aircraft passengers to call correspondents on the ground; this excludes communication required for flying an aircraft by its crew.

The system consists of the following elements:

● 1. Terminals, generally about ten cordless handsets on board the aircraft, to permit passengers to call from their seats.

● 2. On-board equipment to provide control of outgoing calls, particularly queues and 'hand-overs' when the aircraft leaves one coverage zone, defined by a ground antenna, for another zone, and transmission to the ground.

● 3. Each ground station defines a cell and is connected to the public cable network. Distinction is often made between 'en route' stations used when the aircraft are at their cruising altitude between 3000 and 14 000 m, 'proximity' stations which are operational when the aircraft is at a distance of less than 40 km from the airport and at low altitude, and 'airport' stations used when the aircraft is on the ground.

The frequencies allocated to the service are designated L-band and cover the range 1500–1700 MHz. The techniques used for the different systems vary in respect of analogue and digital technology, type of digital coding for speech, access techniques, multiplexing and modulation.

8.3.2 Services and projects

8.3.2.1 *The United States*

The first commercial aircraft telephone service started in the United States in 1984 with the GTE-Airfone company. Initially offered on long haul scheduled flights, the service has been extended to medium range scheduled flights of the principal American airlines. At the start of 1991, 1400 aircraft were equipped, that is one third of the American continental airline fleet, after only six years.

Taking account of this rapid success and to stimulate development, the Federal Communication Commission, the statutory American organisation for telecommunication services, broadcasting and frequency co-ordination, granted four new licences in January 1991 to specialised companies to provide a service similar to that of GTE-Airfone. Each licence holder is contracted to provide an operational service from a minimum of 25 towns in three years and 50 towns in five years.

Two consequences of increased competition can be foreseen. Firstly, a reduction in the price of the service; GTE-Airfone currently invoice £1.40 for establishing a national call and £1.40 per minute of communication, that is £15.40 for a six minute call. An international call is invoiced at twice this. There is thus a ratio of 1 to 8 between a telephone call by aircraft and an intercity call to the United States. Competition between several operators should reduce this difference. Secondly, new FCC licence holders, particularly the In-Flight Phone Company, propose to offer a digital service which is economical in frequencies and permits transmission from on-board personal computers and access to information services.

8.3.2.2   Europe

In Europe, the projects are on the whole less advanced. British Airways was the first airline to equip its aircraft with on-board telephones in 1986. Judged to be too costly, the service has been continued only on some transatlantic routes. In contrast, it has contributed to the creation of a group of airlines, European telecommunications operators and manufacturers who, within ETSI group RES 5, have formulated the European Standard Terrestrial Flight Telephone System (TFTS).

TFTS consists of a direct cellular radio communication system with aircraft using ground stations connected to the public network. At first, the system will be limited to the public telephone before facilities for data transmission and facsimile are offered. In terms of frequencies, 2×1 MHz have been allocated in L-band (1600 MHz) for the experimental phase. The frequency plan foresees requirements of 2×5 MHz. Coverage of Western Europe requires about 50 cells, that is 50 ground stations spaced at around 400 km which permit connection to the public network via concentrating and control equipment. The first tests took place in 1991 and allowed for the possibility of a commercial start in 1992.

Part 3

THE EUROPEAN POCKET TELEPHONE

9 THE PCS OR POCKET TELEPHONE

9.1 THE PCS CONCEPT

9.1.1 Definition of PCS

The concept of a 'Personal Communication Service' or PCS is defined as a mobile telephone service with an essentially urban or suburban coverage characterised by low cost pocket terminals, communications at a price comparable to a cable telephone, and distribution of the services and products to the general public.

This definition is independent of the technology used. It combines the ultimate objective of mobile communication, a service for the general public which is synonymous with competition and low cost and is associated with a person, not a place or vehicle. This highly ambitious objective draws attention to the problems and feasibility of PCS and the progressive evolution towards such a service.

9.1.2 The problem of PCS

This can be simply stated: what standard is most appropriate to the PCS service? There are numerous contending standards: GSM, the European basis of digital cellular; PCN, the British derivative of GSM at 1800 MHz; the American D-AMPS cellular telephone standards and their Japanese derivatives at 900 MHz and 1500 MHz; the European digital cordless telephone standards DECT; CT2 and their derivatives and American equivalents which are being defined.

After the CT2 and PCN projects initiated in the United Kingdom and deployed in Europe, the united States is the scene of intense debate concerning PCS between the various protagonists who support their own technology and the FCC, the arbitrator responsible for frequency allocation.

The dilemma lies in two complementary approaches which only partially attain the objective. Although a PCS standard based on cellular radio-telephone technology is convenient in terms of functionality and extended coverage, its long-term cost could be too high to achieve terminal and communication prices which are compatible with the PCS objective. Conversely, a PCS standard based on a cordless telephone standard is suitable in respect of terminal and communication costs but imposes coverage limits which are prejudicial to willing acceptance of the service.

Progressive evolution will thus be necessary to permit the technologies used to be validated, the different standards and services to be put into practice and finally to let the user choose an adequate service or services.

Standards for radio paging, private radio communication, marine telephony, aircraft telephones and satellite communication have been identified. Radio paging is evolving at the same time in an autonomous manner to become one of the functions of PCS, integrated with the service and the terminals. The competition between private and cellular radio has been analysed in Section 5.3.4. Some sections of the private market will be offered a PCS service. Others, such as conventional PMR and private trunks, are supported by standards specifically adapted to this type of use. This applies also to marine, aircraft and satellite communication.

9.2   PROJECTS

9.2.1   Europe

9.2.1.1   Telepoint

The British telepoint   In 1985, the DTI and several British manufacturers undertook a feasibility study on radio communication using the FDMA transmission technique between a fixed terminal point and a portable handset. The objective of the project was the development of a generation of cordless telephones using digital technology to eliminate the problems inherent in the analogue range. The increase of the cordless total, associated with the low price level, has caused frequency saturation and this implies degradation of communication quality.

Three years later, technical feasibility was validated with the appearance of the first telephones using CT2 (Cordless Telephone generation 2) technology. However, the excessive costs of these digital products could not compete directly with analogue ones. In contrast, the new technology opened the field to new applications. The handset could be used from several bases or terminal points, not only at home or in offices equipped with a cordless PABX, but also in the street near a public terminal point. Market surveys were then carried out and showed the interest of the public in a handset with these three uses.

In 1988, the DTI (the British Department of Trade and Industry) launched an offer to tender for operating licences for telepoint networks using CT2 technology. Several factors were specified:

- 1. Licences are granted for a period of twelve years.

- 2. Interconnection of the telepoint system is provided by two telephone operators, Mercury and British Telecom.

- 3. Terminal points are to be installed throughout the United Kingdom and radio transmissions will use the 864–868 MHz frequency band.

- 4. The telepoint service is limited to outgoing calls. Incoming calls and 'hand-over' are not authorised.

- 5. Use of systems to interim specifications is possible if necessary (the CT2/CAI specifications were not completely defined when the operators were chosen by the DTI) while guaranteeing eventual installation of a system conforming to the definitive CAI specification and providing the possibility of 'roaming' between operators (using a given terminal on a network with the ability to access competing networks).

- 6. An exclusive undertaking between an operator and the proprietor of permitted terminal installation sites is prohibited.

In January 1989, from eleven candidates the following four consortia were retained by the DTI:

- Zonephone, controlled by Ferranti (60%), a venture funding company (31%) and Telephone Rentals (7%) a subsidiary of the Cable and Wireless (C & W) group.

- Phonepoint, combining the operators British Telecom (49%), STC (10%), Nynex (10%), France Télécom (10%) and Telekom (10%).

- Callpoint, consisting of Mercury (40%) of the C & W group, Motorola (30%) and Shaye (30%).

- BYPS, managed by the Barclay (30%), Shell (30%) and Philips (40%) groups.

Three operators opened their service between 1988 and 1989, with only a few thousand subscribers. The reasons for this setback, described in Chapter 7.4.2, do not foreshadow development of projects of the PCS type. They merely show that the first years of launching are very difficult. It is not sufficient to open a new mobile service only partially in order to reap the benefits.

It was necessary to wait four years, until the end of 1992, for the official commercial opening of the only remaining Telepoint operator, Hutchison, which bought the BYPS company and operates on the stable European CT2 standard.

The German project: Birdie In the manner of the English consortium BYPS which called its service Rabbit ('to chatter'), Telekom has chosen the name Birdie (a golfing term signifying 'directly to the objective') for its telepoint service. Two technologies are presently under test: CT1+ at Münster and CT2 at Munich.

The CT1+ system provided by Ascom/Autophon has been operating at Münster since October 1990 with 150 public terminal points and 1500 subscribers. Market studies are currently in progress, and on the basis of the published results of two pilot sites Telekom reserves the right to choose one or the other technology or simply to terminate the experiment.

The initial Telekom project was a decision on the approach to Birdie in mid-1991, but the delay in the marketing of products conforming to the CT2/CAI standard postponed this date to the start of 1992. Initially intended to open in August 1990 with 200 public terminal points and 2000 subscribers the CT2 system controlled by GPT, the British subsidiary of Siemens, was finally put into service in May 1992, more than a year late.

The French project: Pointel Unlike Germany, France does not have competing standards. The choice of CT2 arose from initial technical studies in 1987 to demonstrate to the military authorities that use of the CT2 frequency band would not disturb military systems. Technical experiments were carried out at Valence by France Télécom, Crouzet and Shaye and the military authorities, to obtain the green light for the liberation of frequencies at the start of 1990, the date of launching the call for tenders.

Three calls for CT2 tenders were launched before the opening of a telepoint or 'Pointel' service at Strasbourg in the autumn of 1991 with the commercial name 'Bibop'.

The consortium retained to provide the system combined Cap Sesa, Electronique Mecelec, Sextant Avionique and the coordinator was Electronique Dassault in co-operation with GPT. The manufacturers responsible for portable terminals and private bases are Matra Communication, Sagem and Electronique Dassault. The third project, led by SAT in collaboration with STC, concerns the development of a CT2 cordless PABX connected to the Pointel network.

The development timetable of Bibop has been followed from the start, with the technical and then commercial opening in Strasbourg in 1991 with 250 terminals and 3000 subscribers. The service in Paris then opened at the end of 1992 with a very high coverage rate, since 1500 terminal points have been deployed in the central region of Paris. The subsequent development timetable for other major French cities will be published when the impact of the Bibop service in Paris and its surroundings on the market has been studied.

9.2.1.2 'Personal Communication Networks' or PCN

In February 1989, the DTI, the British Department of Trade and Industry, announced the concept of 'Phones on the Move'. The objective is clear: to create rivalry between the main participants in order to develop a mass radiotelephone market during the next decade. Today, the penetration rate of the only cellular services in the United Kingdom is of the order of 2%. With PCN, analogue cellular networks and GSM, the objective is to attain a penetration rate in the year 2000 of the order of 15%, that is 10 million subscribers.

Discussion of 'Phones on the Move' has brought out a number of service concepts mainly based on the European GSM and DECT standards. It was followed in June 1989 by a call for tenders to select several PCN operators. From around ten candidates, three consortia have been retained.

How can PCN be defined? It is the British version of PCS based on the European cellular standard GSM. It aims to extend the latter from the professional domain to provide a service and product for the general public as with the customary corded telephone.

What are the technical bases? Other than the GSM standard at 1800 MHz, called DCS 1800 and standardised by ETSI in 1990, and a wide band of frequencies of 2×75 MHz, it consists of a microcellular network permitting very high densities of subscribers, 100 000 per km^2, or a user every 3 m! Furthermore, optimisation of the network is envisaged for majority use of the pocket telephone and a reduction of the operating costs of the infrastructures, by using radio beams between radio base and switching stations.

What, therefore, are the principal assets of PCN? The principal strong points are its wide frequency band, high density coverage of subscribers in urban and suburban areas, long-term competitiveness in the cost of services, the sharing of infrastructure (an important factor in the rapid deployment of networks), the quality of the digital radio service and the offer of several value added services.

9.2.2 The United States

PCS in the United States is the subject of numerous often confused discussions and complex processes involving the statutory and legislative authorities, the cable and cellular operators, manufacturers and numerous entrepreneurs. The stakes are high, up to 50% penetration of homes in the year 2000 according to the most optimistic experts.

The development of PCS poses three questions: Which frequency allocation? Which technology to choose and finally which frequency allocation process to set up?

The PCS spectrum can be allocated in three ways.

- 1. The complete liberation of the frequencies (1750–2000 MHz) used by the government for 'experimentation and the emergence of new technologies' has obtained the agreement of the House of Representatives but has been rejected by the Senate and this delays the possibility of such an allocation by a minimum of seven years.

- 2. The sharing of a frequency band at 1850–1990 MHz, which is partly used for other government and civil applications, can be envisaged with a delay of three years in the majority of American towns.

- 3. The third specific approach, which does not have a European equivalent, consists of allocating frequencies without specifying the use, by reserving them for specific private or public use by means of licences. The transmission power of terminals must remain sufficiently low as not to cause interference with other users; this regulation is called 'Part 15'. Although this last solution seems attractive (PCS often uses powers less than 10 mW) its feasibility remains to be proved. However installation would be immediate.

The second problem posed by PCS lies in the choice of technology, a problem common to Europe even if the standards are partly different. Supporters and opponents of the different standards and access techniques confront each other in technical debates which will only be resolved by real experiments on a wide scale. The spread spectrum technique is particularly recommended by supporters of 'Part 15' applied to PCS.

Finally, the allocation of licences, a privilege of the Federal Communication Commission (FCC), is the subject of open debate where four points of view dominate:

- 1. Cellular radiotelephone operators consider that PCS is only an extension of their service to the general public. They consider, therefore, that PCS should be offered on the frequencies which have already been allocated to them by guaranteeing interworking between operators.

- 2. Local telephone companies see PCS as an additional function of the basic network which should be integrated with it. They argue, therefore, for the possibility of offering this new service in an exclusive manner as with their basic cabled service.

- 3. Long distance telephone companies hope, with PCS, to find a new way of by-passing local companies. They argue, therefore, for non-regulation of this service leaving the field free for them to operate on local operators' territory.

- 4. New entrants see PCS as a competitor to the cellular radiotelephone and seek a low cost position for it. The provisional frequency allocation by the FCC gives them the opportunity to test and deploy low capacity systems (with several thousands of subscribers).

The result is a complex situation where doubtless many standards will be evaluated and tested. In the short term this will give rise to the deployment in large towns of incompatible PCS networks operating in 'Part 15' mode; in the medium term frequency bands will be shared and in the long term frequency bands will become dedicated.

9.2.3 South-East Asia

The approach to PCS in South-East Asia is ambiguous. Although there are no projects with PCS operators based on cellular technology in South-East Asia, manufacturers are investing in the development of GSM and DCS 1800 terminals. This is international competition which the European markets should be open to.

The liberal policy of the United Kingdom and the British PCN project will offer, principally Japanese, manufacturers the opportunity to enter the market which is today reserved for GSM by leaving the task of specifying the standard and resolving the problems of the first years to European manufacturers.

In contrast, numerous operators in South-East Asia have shown a lively interest in CT2; Singapore has even gone as far as signing a form of agreement to become a member of the CT2 operators group and thus have access to all the documentation of this standard.

These different countries of South-East Asia—Singapore, Hong Kong, Taiwan and Malaysia—have an ambiguous approach. Are they really interested in deploying the CT2 telepoint service on their own territory or does this interest not reflect the adoption of a position with a view to attacking European Community markets?

An industrial battle for the European pocket terminal market has effectively been declared, and the assets of South-East Asia, such as access to components and the low cost of their labour, will be determining factors.

9.3 STANDARDS, SERVICES AND PRODUCTS

9.3.1 Timescale

There are many criteria for locating services and products. Among these are coverage of the service, the functions offered, the quality of the service, the level of mobility and the price. The coverage of the service and the functions are two different criteria as is the price factor. The coverage of the service can be divided into three categories: 'monosite' for a single fixed location, 'multizone' for multiple and discontinuous geographical locations, 'extended' for a national or international coverage.

The functions of the service can also be divided into three classes: 'general public', with services limited to transmission and reception of calls together with some convenience functions; 'professional, bottom-of-range', corresponding to a service enriched in terms of convenience functions at the terminal level and additional services at network level: 'professional, top-of-range', including the whole range of value added services of messaging and data transmission.

In 1991 in Europe, the services and products offered were limited to the analogue cordless telephone with 'monosite' coverage and functions of the 'general public' type, and the cellular radiotelephone which has extended coverage and 'professional, bottom-of-range' functions.

In 1993, new services and products will be offered. CT2 in its telepoint application corresponds to a 'multizone' service with 'general public' or 'professional bottom-of-range' functions. As for GSM, it will increase the 'extended' coverage of the radiotelephone to international level and functions to the 'professional, top-of-range' level.

In 1995, the services and products offered will multiply while partly segmenting and competing. PCN, initially competing with GSM, is aimed at the general public market with a less extended coverage than the cellular radiotelephone. CT2 will evolve from its initial telepoint application to a more 'general public/monosite' position competing with analogue cordless devices. In the professional domain, the DECT and CT2 standards will compete in offering private cordless systems covering mono- and multi-sites.

Fig. 9.1 shows the chronological development described above, showing the increasing complexity of the relative positions of services and products.

9.3.2 Technical comparison of standards

To compare the various European PCS standards, it is necessary to take account of the technology used, the call transmission and reception

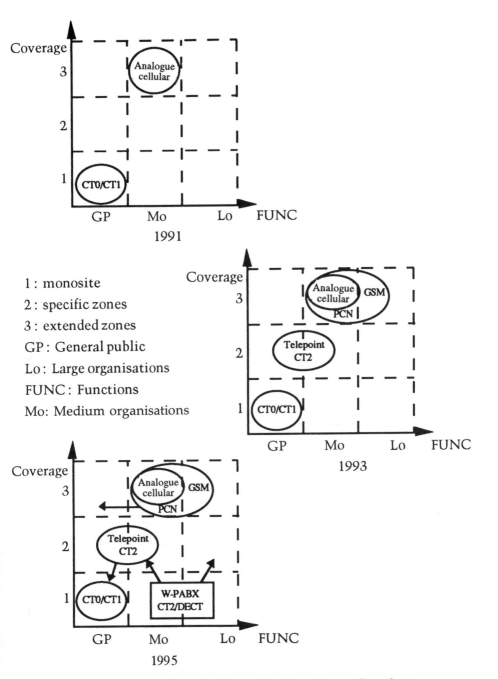

Figure 9.1 Chronological relationship of services and products

capacity, the provision of data transmission services, the level of confidentiality of communications, the reduction of multipath interference (equalisation), the range of radio transmitters which defines the size of monosite coverage areas, the maximum number of bidirectional communication channels per operator and the maximum traffic density supported. These principal criteria are summarised in Fig. 9.2.

Digital cordless telephone standards are thus partly limited in respect of their use in call reception mode and certainly in range. On the other hand, they permit a high density of users per km^2 to be supported. In contrast the digital cellular standards offer a long range but a lower density of subscribers except in the optimised case of PCN/DCS 1800 in an urban area where the size of cells is greatly reduced.

Standard	CT0/CT1	Analogue cellular	CT2	GSM	PCN/DCS 1800	DECT
Frequencies (MHz)	16/47 26/41 900	200 400 900	900	900	1800	1900
Technology	Analogue	Analogue	Digital	Digital	Digital	Digital
Incoming/outgoing call from terminal	Yes	Yes	Yes[1]	Yes	Yes	Yes[1]
Data transmission	No	Limited	Yes	Yes	Yes	Yes
Confidentiality	Limited	Limited	Yes	Yes	Yes	Yes
Equalisation	NA	No	NA	Yes	Yes	NA
Frequency hopping	No	No	No	Yes	Yes	No
Maximum transmitter range	200 m	35 km	200 m	35 km	7 km	200 m
Maximum number of channels per transmitter	8/15/40	24	40	72[2]	144	144
Spectral efficiency in Erlangs/MHz/km^2	NA	5 (range 1 km)	50	40 (range 500 m)	160(range 250 m)	500

[1] The call function is activated manually.
[2] Re-use factor of 7.

Figure 9.2 Technical comparison of standards

Standard	CT0/CT1	Analogue cellular	CT2	GSM	PCN/DCS1800	DECT
Network services	Incoming/outgoing calls	Incoming/outgoing calls, call transfer, call barring	Outgoing call, manual activation of the incoming call function	Incoming/outgoing calls, supplementary services, data transmission	Incoming/outgoing calls, supplementary services, data transmission	Outgoing call, manual activation of the incoming call function, hand-over
Terminal convenience functions	Memory + Last number recall	Hands free, alphanumeric directory, DTMF	Alphanumeric directory, pocket telephone	Hands free, alphanumeric directory, answering machine, voice-activated dialling	Optimisation for portable use	Voice and data functions, PABX type service
Launch of service	1970	1982/1985	1991	1991	1993	1993
Type of marketing	General public	Service provider, radio installers, car accessory suppliers	Operators, professional and consumer distributors	Operators, Service providers, radio and car installers	Professional and consumer distributors.	Telephone installers.
Price	£100 (terminal)	£100/month (service + terminal)	£20/month (service + terminal)	£90/month (service + terminal)	£50/month (service + terminal)	£400 (terminal)

Figure 9.3 Comparison of functions, standards, services and products

9.3.3   Comparison of functions, standards, services and products

To complete the analysis, the functions offered by the networks and terminals, the dates of entry to service, the type of marketing and distribution, and the price level of the service and the terminals will be compared. Fig. 9.3 summarises the various points.

In terms of price, the CT2 telepoint is the lowest cost service, less than half as dear as its cellular competitor the PCN/DCS 1800. DECT has not been included here as a telepoint service although it has the capacity, but as a standard for professional application as a cordless company telephone system. Distribution certainly conforms to these different positions and shows an evolution towards distribution to the general public of all these products.

10 EVOLUTION OF THE PCS

10.1 THE SCENARIOS

10.1.1 Market uncertainties

The projects which specify new European radiocommunication services for the next ten years are entering their final phases. Each of the three classes of mobile service—cellular, cordless and radio paging—is being considered by dedicated ETSI standardisation committees and its specifications should shortly be established. Technical developments will then be minor. In 1991 the GSM and DCS 1800 cellular, DECT and CT2 cordless, and ERMES radio paging standards were published.

However, although the technical facilities offered by these new personal radiocommunication systems (PCS) are already defined, marketing uncertainties remain. The PCS, whose concept was described in Chapter 9.1, is aimed at a general public market with extended urban coverage. Cellular systems, as well as cordless ones for telepoint application, can meet this objective. Henceforth, two questions arise for operators and European manufacturers involved in the future of personal communications.

- 1. The first concerns the transition of a market, which is now essentially professional, to the general public. The pocket telephone satisfies a universal requirement of people while travelling, but will the private consumer accept the additional tariff charges of subscriptions and mobile communications in comparison with a basic corded service?

- 2. The second question concerns the division between telepoint and cellular services. Will these mobile services eventually become directly competitive with the possible disappearance of one of them, or will they remain complementary as they are today?

Three basic scenarios in the evolution of PCS can be foreseen, according to the tariff policy of the operators, technological developments by manufacturers and the reception by the general public market.

10.1.2 Contrasted scenarios

The three assumptions for the development of PCS are in contrast and depend on several criteria.

- 1. The 'Professional PCS' scenario predicts the commercial failure of general public telepoint and cellular approaches. Mobile services will remain limited to professional use with stagnant penetration rates below 10%. The general public PCS will not exist.

- 2. This scenario describes the success of the 'General public PCS' specified in two parts. The first shows cellular technology evolving towards a low range approach using cordless services, while the second restricts the offer of a general public PCS to cordless services only. The PCS terminal is a pocket telephone of the radiotelephone or cordless telephone type.

- 3. The 'multiproduct PCS' scenario presents the continuation of general public cellular and cordless services with a range of products specified according to use. The user's single subscription permits access to all the PCS services via a multiplicity of products.

10.1.3 The criteria

The criteria justifying these scenarios concern the statutory policy, the extent of exposure of operators to competition, the strategy of these operators and manufacturers together with the behaviour of users. But the factor which determines success remains a price of products and mobile services which is acceptable to the general public. The criteria defined below are presented in Fig. 10.1 for each scenario.

Type of service	Annual threshold of acceptability (product + service)
Cellular	£600
Telepoint	£350

Figure 10.1 Threshold of acceptability by the general public

- 1. The only psychological and behavioural criterion is the possible reluctance of the consumer to use a pocket telephone. He can feel deprived of some of his liberty by being able to be contacted continuously and harassed by this new personal product which connects him to an invisible network.

- 2. The statutory factor determines the level of competition between operators in terms of the number of PCS licences allocated and hence, indirectly, the subscription and communication tariff policy.

- 3. The PCS operators' tariff strategy will be determined by the choice of scenario. The current profitability of the various cellular networks is very high. The annual turnover generated by an average cellular subscription is equal to the initial investment of the operator for installing one cellular subscriber line. A typical case is that of the British operator Racal Vodaphone presenting results of the order of 40% before taxes with communication costs of 30p per unit. The operator can thus foresee a fall of tariffs by 30% while maintaining acceptable profits. Opening to competition will initiate a fall of tariffs, but only a willingness of the operators to develop a market larger than a purely professional one will permit the threshold of acceptability to the general public to be attained.

- 4. The public price level of the products will depend essentially on the economies of scale achieved by the manufacturers. Experience of this type of activity shows a division of fabrication costs by approximately two when the cumulative quantities are multiplied by ten. This rule must be modified if the provision of specific components is not controlled, as was the case in the computer industry with dynamic memories. In this case, the component manufacturers artificially created a shortage in order to cause an increase in the price of these components.

- 5. Nevertheless, the threshold of acceptability by the general public remains the key factor in the take-off of the market. In fact, there are well defined price barriers for all marketed products. As long as the price of answering machines and cordless telephones was above £200 in France, the market remained relatively limited; it started only when the price fell below this barrier. Similarly, the general public threshold of acceptability of camcorders is estimated to be £800.

  This criterion relates to the total annual cost of using the mobile service. It includes the price of the product amortised over its lifetime, the annual subscription and the cost of mobile communication. Market surveys give mean use of communication, lifetimes of the order of three years for the products and estimates of the threshold of acceptability of the annual cost. Fig. 10.1 gives the price objectives of cellular and telepoint systems to permit the general public market to be attained.

- 6. A final criterion to take into account concerns the fiscal aspect. It could be tempting to tax the possessors of telephones with a specific tax similar to car tax and television licences. This new tax has actually been proposed in the United Kingdom and in Italy and could extend to all of the European Community in a few years. The temptation for public bodies to use the PCS subscription as a new source of fiscal income to complement that from the car driver and television viewer is high. The level of this tax could be highly dissuasive when considering a PCS subscription.

10.1.4 The professional PCS

The first scenario assumes failure to penetrate the general public market by mobile services. Mobile communication remains limited to professional use with stagnation of the penetration rate below 10%. PCS, which aims to achieve a mass market, does not exist. Mobile telephone activity eventually becomes comparable to that of other professional products like the PABX, professional video and sound systems and portable microcomputers.

This commercial block is explained by the reluctance of the general public to be able to telephone and be called at any time and also by the high cost of the products and services.

In fact, the complexity of the technologies used does not permit a sufficiently rapid decrease of the costs of these products. Mobile systems, like the associated terminals, remain at a high price which is prohibitive for achieving the limits set by the general public. The cost of the service remains equally high. Mobile systems require high investment but, above all, the objectives of operator profitability remain such that the targeted subscribers must be from the professional categories in order to support the annual cost of the service (subscription and communication) of £350 for the telepoint service and £700 for the cellular one. The private user, who consumes less than a professional, is less profitable to the operator. The marketing and communication policy of the service will remain essentially professional.

The cellular and cordless markets co-exist. The radiotelephone is tending towards an extended regional or national coverage while cordless systems remain limited to local urban coverage. DCS 1800 is being considered as a cellular service on another band of frequencies to permit the problems of network saturation in regions of high subscriber density to be avoided.

10.1.5 The general public PCS

10.1.5.1 A mass market

This approach presents the success of mobile services with the general public. The pocket telephone is a reality for all, and the operators control tens of millions of subscriptions in Europe. The costs of products and services, profiting from the effect of volume and the developments of new technologies, have been greatly lowered to attain the threshold of acceptability of the general public. This scenario divides into two cases depending on the technology used for the new pocket telephone—cellular or cordless.

10.1.5.2   The general public cellular PCS

This case envisages the advent of the radiotelephone, an approach suited to the professional market when offered with top-of-range services, and to the general public market when offered with simplified services. The initial objective of the British PCN is attained.

This dual suitability supplants the market for the cordless telephone in its telepoint application. This service will only have survived for a while as a hybrid interim solution to allow time for the technology of the radiotelephone to become simpler and its costs lower. The success of the cellular approach will thus logically lead to the disappearance of the cordless telepoint application. The cordless telephone remains at the house or office.

Specific low cost subscriptions are offered to the general public in urban areas with microcellular systems supporting densities of one subscriber every 3 m, while dearer subscriptions offering national coverage with macrocells remain oriented towards professionals.

10.1.5.3   The general public cordless PCS

This case, a corollary of the previous one, divides mobile services into two distinct markets. Cellular systems remain complex with a high cost of use and are reserved for professionals while the telepoint cordless approach is aimed at the general public.

The British PCN has failed with respect to its initial objective. The objective of multiplication of DCS 1800 microcells to achieve high subscriber densities is irreconcilable with the reduction in price of the service, because of the increase of the cost of the system per subscriber. DCS 1800 remains as an additional cellular network, the only difference being the use of a higher range of frequencies.

A high density of public telepoint terminal points covers the majority of urban centres together with the service areas of the major European trunk roads. It is possible to telephone from these commercial centres, stations, and airports without having to search for the logo of the telepoint operator. The installation of terminal points is such that transmission regions overlap to give a continuous coverage.

The pocket telephone can be used at home, in the office and in the street and has become a consumer product in the same way as a television or a portable radio.

10.1.6 The multiproduct PCS

10.1.6.1 The varied requirements

This scenario presents the difference between the theoretical concept of a universal pocket telephone which satisfies all the situations of everyday life and the reality of the requirements. The one single product does not exist and does not have a meaning. The PCS, a true mobile service for the general public, exists but is supported by a complete range of mobile telephones, each responding to a precise need.

10.1.6.2 A range of products

The natural division of requirements does not lead to the use of a universal product, but rather to the development of a range of multistandard products. As the evolution of radio sets divided, according to requirements, into car radios, portables, high fidelity systems and even headsets in aircraft, the mobile telephone will follow a similar progression into the car telephone, pocket telephone, domestic cordless telephone and mobile telephone in aircraft and trains. This comparison is reproduced in Fig. 10.2.

10.1.6.3 A range of services

The range of products corresponds to a range of services for the conditions of use with a single subscription. Fig. 10.3 presents the correspondence between the products, the services and the subscription.

PCS telephone	Radio products
Car telephone	Car radio
Pocket telephone	Portable radio
Multifunction cordless telephone	Domestic high fidelity system
Aircraft telephone	Radio headphones in aircraft and trains

Figure 10.2 Equivalence of PCS and radio product ranges

Situation	Multiproduct	Multiservice	Subscription
Vehicle	Car telephone	GSM 900 cellular	
Pedestrian	GSM pocket telephone	GSM900/ DCS 1800 cellular	SINGLE SUBSCRIPTION
Pedestrian	Telepoint pocket telephone	Telepoint	TIM-SIM
Aircraft/train	Telephone in aircraft and trains	TFTS (Terrestrial Flight Telephone Service)	

Figure 10.3 The correspondence of products, services and subscriptions

On taking out a subscription, the user specifies his conditions of use and the desired additional services which are then activated and defined in a single memory card (SIM) which is independent of the equipment.

Extending the Subscriber Identification Module (SIM) card to all mobile terminals permits the physical product to be decoupled from the service. In practice, this memory card contains all the data relating to the subscription and its owner. The standardised terminal can be sold by a distributor to the general public without a card as a conventional electronic product (such as a radio, television or camera). The card is obtained from the operator or an approved distribution network with the chosen subscription data and the number of the account to be debited. It is then merely necessary to insert this SIM card into the telephone in order to use it. The innovation of the SIM thus increases the 'pay-per-call' possibilities. The subscriber pays in accordance with the use which he makes of his pocket telephone and not on the basis of a complete subscription at high cost.

A variant of this scenario is the eventual reduction of the number of portable products by combining the cellular and telepoint functions in the same pocket terminal.

10.1.7 Summary

Fig. 10.4 recalls the significant criteria such as the tariff policy of the operators, the product price level, and the type of distribution of PCS products and services for the scenarios envisaged.

Scenario Criteria	PCS Professional	PCS General Public		PCS Multiproduct
		Cellular	Telepoint	
Statutory	T : low	T : low	T : high	T : high
Number of operators	C : low	C : high	C : low	C : high
Tariff	T : high	T : high	T : low	T : low
Price of communication	C : high	C : low	C : high	C : low
Terminal price	T : high	T : high	T : low	Segmentation
	C : high	C : low	C : high	of prices
Type of PCS	PCS does	Cellular	Telepoint	Mixed
service	not exist	Service	Service	C and T
Penetration rate	T < 10%	T < 5%	T > 20%	T > 15%
	C < 10%	C > 20%	C < 5%	C > 15%

C : Cellular - T : Telepoint.

Figure 10.4 PCS scenarios for the year 2000

10.2 POINT OF VIEW

10.2.1 'Not written in the stars'

As declared by Mr J. Carey and cited at the start of this work: 'The communications landscape of the next ten years is not written in the stars'. In fact, the current strategy of operators and manufacturers in this sector, together with the policy of the European Commission, will be decisive in the future of the pocket telephone.

All the scenarios described above are probable, but the increasing importance given to the mobile radio communication sector by the various participants tends towards the third scenario, that of 'multiproduct PCS'. This evolution should have three distinct phases up to the year 2000, in which each participant will have a decisive role. The launch phase between 1990 and 1993 will be dominated by the prominence of national public operators. The second phase from 1994 to 1996 will be the transition period directed towards a reduction of terminal price and associated with competition between manufacturers and the effect of volume. This phase will also see the entry of private operators which will stimulate competition between operators and lead to a lowering of the service cost (subscription and communication). During the third phase from 1997 to the year 2000, the reduction of the price of PCS products and services really will permit entry to a mature phase where distribution to the general public fully plays its part; the PCS telephone will then be a consumer product in its own right.

10.2.2 The launch phase

The launch phase has already started across the whole of Europe with the GSM and CT2 projects. It will spread from 1990 to 1993 with much standardisation and development work on the part of national public operators and European manufacturers.

Large orders for several hundred thousand terminals will be allocated to manufacturers in order to initiate rapid development and lower the price of digital mobile terminals. But the stakes will also be political; the initial choice of manufacturers will determine the dominant European manufacturers of pocket telephones. In fact, this product remains one of the last bastions of European consumer electronics in the face of the Japanese offensive.

In parallel, the magnitude of development costs leads to partnership strategies on the part of manufacturers. The time is no longer one of national standards; the markets are immediately international. Sharing of development costs and common use of production facilities to achieve economies of scale have become indispensable.

It is during this phase that the policy of the operators will be dominant and will direct the manufacturing policies of the sector.

10.2.3 The transition phase

This period, from 1993 to 1996, will see a decrease of the price of digital terminals, due to the economies of scale realised during the previous period with large orders from public operators, and because of the arrival of Asian competition, notably Japanese and Korean.

In fact, around ten Asian manufacturers are developing ranges of digital mobile terminals to the European standard. 1993 is the effective opening of the single market, and also the period of deployment of digital mobile networks. Their objectives of taking sections of the market will consequently lead to a very aggressive commercial policy in order to erode the European manufacturers' share of the sector.

Service costs will also be lower, but to a lesser extent due to the preponderance of monopolies. Service competition will not really be involved until several years after the opening of the GSM and telepoint networks.

10.2.4 The mature phase

It is in this third period from 1997 to the year 2000 that the combination of lower product and service prices will permit the general public thresholds of acceptability to be achieved. Commercial distribution policy will thus play a major role in spreading the PCS concept widely.

At these times the networks will be consolidated to accept a very large number of subscribers and competition between operators for the range of services offered will fully play its role of catalyst; multistandard multi-products will thus be available to satisfy all subscribers' requirements and the pocket telephone will be a reality.

11 THE EUROPEAN POCKET TELEPHONE

11.1 THE INTERDEPENDENCE OF FOUR KEY FACTORS

Analysis of the various domains of radiocommunication and the evolution of this sector towards the PCS leads to the identification of four key factors which are fundamental to this sector and its evolution. They are the statutory and European authorities, the operators, the system and terminal manufacturers, and the service and terminal distributors.

Their interdependence is very strong. All participate in the work of standardisation and structuring of the radiocommunication sector at European level. Each successively plays a leading role in the development phases of PCS; these are the manufacturers in the design phase, and the operators in the service launching and terminal initialisation phase (the set of terminals is large). The terminal manufacturers and new operators contribute to a price reduction in the mature period by means of economies of scale in the products and competition in the service, and distributors ensure a combination of commercial presence and added value of the products and service.

Although each needs to play its role, the economy of the sector depends on a balance between the parties and the British market has demonstrated this to an absurd extreme. In practice, there is price uniformity between terminals and communication. By greatly subsidising the purchase of cellular terminals, the British operators and their distributors have created a fictitious mobile telephone market which is neither stable nor 'consumer' and is incapable of paying for communication (see Section 6.2.3.2). They have artificially lowered the price of terminals, involved some distributors with few scruples of healthy management in loss leaders and created abnormal effects of user turnover. Finally, they have contributed an unbalance of margins on the added value of the sector which are very much in favour of the distributors. They have favoured control of the terminal market by American and Japanese companies without any concern for taking part of the market at the expense of European manufacturers.

This extra-European domination in certain terminal markets is a menace for the whole sector and it is not limited to European manufacturers. In addition to selling products to distributors, the control of distribution of products and services is intended to change the added value of the network in favour of terminals by integrating a maximum of functions into them. This is the strategy of Motorola in respect of operator services for the sale of systems and terminals, and of the Japanese and Koreans in the distribution of terminals. The European standardisation organisations are no longer unaware of these menaces; their long term independence, as a result of the competition which will arise for a world standard PCS, depends on the size of the European market for the manufacturers and operators who form the standardisation authorities.

There is, therefore, a common interest between standardisation organisations, manufacturers, operators and European distributors in the public and private sector.

11.2 THE COMMUNITY APPROACH

11.2.1 A worrying situation

The telecommunication equipment sector, of which the terminal section is regularly increasing, remains one of the rare sectors of consumer electronics where Europe has, until now, maintained a positive commercial balance. On the other hand, Fig. 11.1 shows that the deficits in other information technology activities are increasing regularly.

Although still protected, the European telecommunication equipment sector will experience an unprecedented international offensive from third

Billions of ecus	1985	1988
Data processing	15	17
Components	3	4
Consumer electronics (VCR, Camcorder etc.)	10	12

Source: CEC (88), EIC (89), Butler Cox (90).

Figure 11.1 The deficit of information technology in Europe

party countries. In fact, estimates predict that expenditure on telecommunication equipment will be the largest European investment in the coming years, with a total of 1000 billion ECUs in twenty years.

Protected by the multiplicity of national standards, the penetration of terminal products from Japan is nevertheless accelerating with the appearance of top-of-range telephone products such as facsimile machines and cordless telephones. The share of facsimile machines made in European factories is estimated at only 15%, of which the technology and key components are from Japan. This tendency is confirmed. Fig. 11.2 shows the predicted evolution of sales of telecommunication equipment between 1987 and 1993, and the increasing share of the terminal market.

This evolution is thus worrying in more than one sector, and specifically in the sector of the pocket telephone of the year 2000. Competition with Japan is the most severe in markets which are increasing strongly and being deregulated, such as facsimile machines and radio terminals. European industry has already lost the facsimile battle. Radio terminals, precursors of the future pocket telephone, remain its last chance.

11.2.2 Prior conditions

The objective of the Delors Commission which dates from 1984, was to establish, on 1 January 1993, a single area which is common to 320 million people and is based on the principle of free circulation of goods, services, capital and for the following purposes:

● 1. To provide consumers with new products as a result of technological developments and at the best price.

● 2. To establish a single market to provide the economies of scale necessary for the international competitiveness of European manufacturers.

● 3. To guarantee the investments of network operators who offer new services.

Billions of ecus	Total communication equipment sales	Sales of communication terminals	% of terminals in total sales
1987	21.5	6.7	31%
1993 (estimate)	32.4	11.7	36%

Source: Eurostat, Sagatel, CCE.

Figure 11.2 Sales of telecommunications equipment

This principle has affected the telecommunications domain since the European Commission established the prior conditions for the institution of this single market several years ago. The major points of its policy are as follows.

- 1. To facilitate access to this market for manufacturers and operators by creating European standards and by eliminating national technical and commercial boundaries.

- 2. To create a statutory regime which is favourable to technological innovation and supports the competitiveness of European organisations.

The liberal policy of opening telecommunication products and services to competition has two aspects.

- 1. Judicial actions can be employed against member states which are in breach of European directives.

- 2. Large manufacturing conglomerates can be subjected to thorough investigation in order to determine whether the conditions of free competition can operate. This aspect is carefully examined in the case of mergers or acquisitions of companies, in order to avoid monopolies. Financial support allocated by public bodies is also rigorously examined and controlled.

European jurisprudence has existed for several years on which the various policies can be based. These points are fundamental to the operation of the single market and the decisions taken remain of a supernational order. The European Court of Justice has recently rejected attacks by France and Italy concerning the directive on the liberalisation of telecommunication terminals.

11.2.3 The catalysts

The catalysts of the European telecommunications policy established in 1987 are of several types:

- 1. Statutory, with systematic liberalisation of the terminal equipment and service markets.

- 2. Standards oriented with the creation of the European standardisation institute (ETSI) and European standards.

- 3. Creative, with the RACE and ESPRIT research and development programmes on advanced innovative technology and associated with market innovation.

● 4. Political, with the STAR programme aiming to promote use of telecommunication services in less favourable regions of the European Community.

The telecommunications sector has become a priority aspect of the European Community, whose enterprising actions since 1987 are already bearing fruit.

In the standardisation sector, ETSI has already achieved much since all new radiocommunication services put into operation from 1991 and 1992 in Europe are to a single standard. Procedures for mutual monitoring of agreements together with test and conformity procedures are operational. Services will recognise the change since administrative monopolies have been broken leaving the field free to private enterprises.

The competitive environment of the single telecommunications market is ready and national barriers have been eliminated giving Europe second place with 30% of the world telecommunications market and the advantage of a non-negligible increase.

11.3 THE DISTORTING MIRROR OF THE EUROPEANS

11.3.1 The failure syndrome

The European telecommunication market in 1989 represented nearly 20% of the world market for equipment and services; that is 25 billion ECUs for equipment and 85 billion ECUs for services of which 85% is represented by the basic telephone service. This places Europe second in world ranking, behind the United States (40% of the world market) but a long way ahead of Japan (12%).

Although this market is one of the largest in the world, many European manufacturers are leaving this sector because of the offensive of third party countries. Competition operates on all fronts. In the public switching domain, AT&T has adopted a strong position by setting up in countries such as Italy, Spain and Portugal to expand in a few years from 2.5% to 14% of the European market. Motorola has become the leader in radio-communication terminal systems in numerous European countries. The facsimile market is totally in the hands of the Japanese such as NEC, Matsushita, Toshiba and Mitsubishi. The cellular and cordless radio communication terminal domain is greatly divided in the products offered owing to the multiplicity of standards. New generations of European products will provide an opportunity for concentration of world manufacturing in which Europeans should be able to remain competitive.

In spite of the advantages of the single market, numerous manufacturers are surrendering in the face of this offensive and leaving the telecommunication terminal sector because of the lack of global competitiveness of their products; this is the first step before telecommunication systems and finally services are abandoned. The causes of this lack of competitiveness are clear, but the non-negligible advantages remain. The decline of the terminal industry is not inevitable.

Europe is effectively passing through a period of industrial upheaval and is taking refuge, as the United States and the United Kingdom already have done, in a service economy. Studying the reasons for these failures in the distorting mirror of the past, a certain resignation is evident in witnessing the disappearance of whole sections of the European manufacturing economy such as computers, consumer electronics, cameras and watches.

11.3.2 The weaknesses

Europe has numerous weaknesses in comparison with third party countries but the main point is weak technological and commercial development strategies based on the short and not the long term. This short-sightedness is due to the type of financing used by European companies. The principle of capital contribution by shareholders who wish to obtain the highest possible short-term benefits is not compatible with a strategy of new product development and the conquest of new markets.

Since the initial financial risk is often proportional to the degree of innovation of the product, companies thus have a tendency to minimise innovation and concentrate on research with immediate benefits to mature products. Similarly, the profitability achieved during the conquest of new markets is often very small, if not negative, in the first ten years of establishment. These two factors often destroy innovation by the enterprise to minimise the risks taken. In the life cycle of terminals, which is now reduced to eighteen months, competitiveness arises through innovative functions, design, physical performance (size and weight) and price.

Other than the design, the hardware performance and the price are the keys to the pocket telephone market. The objectives of size and weight (several hundreds of grams and several hundreds of cm^3) can be achieved only with assembly technologies and components which do not exist widely today except in Japan. Price competitiveness also arises from control of the whole process.

With the opening of terminals to competition on the European market, local providers will then be in competition with Japanese manufacturing conglomerates which combine the fabrication of single source key components with command of the most recent assembly technologies. These are still unknown in Europe and are often derived from other consumer products such as the camcorder; a considerable volume of capital is available by borrowing at exceptionally low interest rates. The fear of some companies to engage in an unbalanced combat and take refuge in more protected sectors can be understood.

Europe certainly has its weaknesses, but it nevertheless retains non-negligible advantages on its own territory.

11.3.3 The advantages

Europe has a considerable potential, 25% of the world market for telecommunication equipment and services, and a rate of increase estimated at more than 7% per year (more than 20% for single terminals) without counting the opportunities offered by entering the market economy of Eastern countries.

Although favourable to competition, the establishment of European standards is a decisive advantage, on condition that rules of operation, both internal and external with third party countries, are instituted. Economies of scale can then be fully realised and be profitable to manufacturers, operators and consumers.

Following this logic, public markets remain a priority of industrial competitiveness in terms of the size of orders placed (600 million ECUs in 1987) and the section of the global telecommunications equipment market which it represents (85%).

The European telecommunications industry is also positively placed on the world playing field. Fig. 11.3 presents the concentration of information technology sectors and it can be seen that it is only in the telecommunications sector that European enterprises can retain a dominant role in 1995.

	Consumer electronics (1)		Semiconductors		Telecommunication	
Year	1979	1995	1979	1995	1979	1995
Europe	31.0%	6.8%	10.4%	NS	33.0%	41.0%
United States	6.4%	NS	69.1%	7.8%	63.1%	36.6%
Japan	61.6%	93.2%	20.5%	92.6%	3.9%	22.4%

(1) Other than communication terminals.
NS : Not significant.

Figure 11.3 Telecommunications manufacturing

For Europe to be not only one of the largest markets of the world but also the centre of research, development and production of innovative products, concerted action must be taken by all European participants. Action must be taken from this point of view so that levels of employment and production in Europe should not be reduced.

11.4   A PLANNED EUROPEAN ECONOMY

The stakes involved in PCS are considerable for Europeans. It requires concerted 'Japanese style' operations in the European radiocommunications sector. The initial problem is how to avoid domination by the Asians and Americans in a sector in which the future involves consumer electronics which are not well understood by European manufacturers.

Four types of action are necessary to operate in this context:

- 1. A partnership of operators and manufacturers well beyond the co-operation which leads to the creation of European standards. Public and private European operators and manufacturers must co-operate by combining their research and development resources to design, develop and manufacture competitive products at European level. This would be in the image of the co-operation between NTT, the Japanese public operator, and Japanese manufacturers for the development of leading systems as much by the level of miniaturisation as by the quality and functions offered. The operators contribute by their knowledge of standards, the requirements of users through their activities in providing services and products, and by their researches in support of advanced technique and technology. The manufacturers contribute by their expertise in the development and sophisticated manufacturing of products, and co-operation with the suppliers of specific components. They guarantee price reduction in accordance with the economies of scale expected from the quantities of products arising from distribution by operators.

- 2. Another domain of co-operation between operators and manufacturers relates to distribution. For the PCS this will combine products and services in a complete package for the final user. Operators and manufacturers have the same goal of providing and controlling distribution. The provision of specific products and services specific to the requirements and context of use is a means of achieving this objective.

   An extension of this type of co-operation is the formation of consortia of service operators outside the EEC in countries which introduce public radiocommunication services.

- 3. The third necessity for a balanced increase of European PCS lies in the need to distribute a share of the margin created by the overall added value of the sector. Too great a proportion, as is currently the case in favour of the operators, is prejudicial. It signifies that one part of the margin effectively created by the manufacturers and the distributors is transferred to the operators and/or the negotiating power of the sector is too much in favour of the latter. In this context two developments are possible:

    3.1 A drastic reduction of the operator's profits by the creation via the statutory authorities of very strong competition for services, this is the classic pendulum effect. Very strong competition for services occurs in long distance cable services in the United States and generates much confusion. Services are offered which are often suboptimal since investment is concentrated in regions which promise the highest profitability as opposed to the provision of uniform services.

    3.2 Redistribution of margins in favour of system manufacturers, terminal manufacturers and distributors. This is made possible particularly by the influence, which will increase, of operators in combined terminal service distribution. Although professional cellular radio-telephone distribution is currently performed by 'service providers' which are third party companies independent of the operators, the increasing number of users and the reduction of prices, combined with a willingness of operators to expand abroad, will encourage amalgamation of European operators and 'service providers'. To control the end-user service in Europe, operators must engage in a European policy of service and product distribution.

- 4. The fourth action must aim to establish regulations in respect of non-European manufacturers to guarantee fair operation of the rules of competition. Competitiveness between organisations is largely based on the cost structure. Market share, by creating economies of scale, alters the cost structure and contributes to the process of competition. The cost advantages, associated with the Asian financial market whose interest rates are very low, and with the poor social development in certain far eastern countries, are not often accessible to European companies. True competition on the European market should minimise these advantages. For this to happen, component purchasing, development and product production must be performed in Europe, and in practice this is difficult to control. Under these conditions, a European policy similar to that operated in the European automobile sector is necessary now and for the next ten years.

GLOSSARY OF
ABBREVIATIONS

AMPS   Advanced Mobile Phone Service, the American standard for cellular
            radio telephones.

BSC   Base Station Controller, radio transmitting/receiving station
            controller of the GSM system.

BTS   Base Transceiver Station, radio transmitting and receiving station of
            the GSM system.

CAI   Common Air Interface standard. Applied to the CT2 standard and
            designates the signalling protocol associated with the radio
            interface.

CAMT   World telegraph and telephone administrative conference.

CCITT   International Telegraph and Telephone Consultative Committee.

CCIR   International Radio Communications Consultative Committee.

CDMA   Code Division Multiple Access.

CEN   European Standardisation Committee.

CENELEC   European Electrical Engineering Standardisation Committee.

CEPT   European Conference of Postal and Telecommunications
            Administrations.

CNES   French National Centre for Space Studies.

C-NETZ   German standard and analogue cellular network.

CT0   Cordless Telephone generation 0, analogue cordless telephone
            standard used in the United Kingdom and France with several
            variants.

CT1   Cordless Telephone generation 1, analogue cordless telephone
            standard standardised at the European level under the auspices of
            the CEPT and used particularly in Germany and the Scandinavian
            countries.

CT2	Cordless Telephone generation 2, digital cordless telephone standard standardised by ETSI.
CT3	or DCT900, cordless telephone system developed by the Ericsson company.
D-AMPS	Digital AMPS, American digital cellular standard.
DECT	Digital European Cordless Telecommunication system, digital cordless telephone standard standardised at the European level under the auspices of ETSI.
DRG	Directorate of General Regulations in France.
DSRR	Digital Short Range Radio, European standard for private short range radio communication or 'digital CB'.
DTI	Department of Trade and Industry (UK).
ECTEL	European Committee of Telecommunications and Electronic Professional Industries.
EEC	European Economic Community.
EFTA	European Free Trade Area.
ERMES	European Radio Messaging System, European standard for one-way messaging.
Erlang	Unit of telephone traffic flow.
ETSI	European Telecommunications Standards Institute, European standardisation organisation.
EUTELSAT	European satellite communication organisation.
FDMA	Frequency Division Multiple Access.
GPS	Global Positioning System, world system for radio navigation by satellite.
GSM	Group Special Mobile or Global Service for Mobile Communication, European standard for digital cellular radio telephone.
Hand-over	Intercellular transfer.
INMARSAT	International satellite communication organisation for marine applications.
IFRB	International Frequency Registration Board.
ISDN	Integrated Services Digital Network.
ITU	International Telecommunications Union.
MOBITEX	Swedish standard trunk.
MSC	Message Switching Centre, a switch.

NMT   Nordic Mobile Telephone, Scandinavian analogue cellular standard.

OFTEL   Office of Telecommunications (UK).

PABX   Private Automatic Branch telephone Exchange.

PCN   Personal Communications Network.

PCS   Personal Communications Services.

PMR   Private Mobile Radio, a conventional private network.

POCSAG   Post Office Code Standardisation Advisory Group, radio paging standard.

RACE   Research into Advanced Communications in Europe.

RADIOCOM 2000   French analogue cellular standard and network.

RTMS   Italian analogue cellular standard.

SIT   Telecommunications Industries Syndicate.

SPER   Syndicate of the Professional Electronic and Radio Equipment Industry.

STAR   Special Telecommunications Action for Regional Development.

TACS   Total Access Coverage Service, British analogue cellular standard.

TDMA   Time Division Multiple Access.

TIM   Terminal Identification Module.

Trunk   Private network of shared resources.

UMTS   Universal Mobile Telecommunications System.

WARC   World Administrative Radio Conference.

X.25   Packet data transmission standard.

INDEX